BEYOND INCLUSION

Fintech's AI Inflection Point

Dr. Tadiwanashe Muganyi

First published 2026

ISBN:
Ebook – 978-1-970853-81-0
Paperback – 978-1-970853-82-7
Hardcover – 978-1-970853-83-4

Printed in the United Kingdom

Note on Data and Accuracy

The statistics, figures, and factual claims presented in this book reflect the most recent publicly available data at the time of writing, drawn from official sources including central bank publications, regulatory filings, World Bank datasets, peer-reviewed research, and verified industry reports. Financial markets, regulatory regimes, and technology platforms evolve rapidly, and some figures may have changed between the completion of this manuscript and its publication.

Every effort has been made to ensure factual accuracy. However, in a work of this scope — spanning multiple jurisdictions, institutions, and rapidly evolving fields — errors in detail may persist despite rigorous verification. Where they exist, they are the author's own and don't diminish the central thesis of this book: that the convergence of fintech and artificial intelligence amounts to a structural inflection point in global financial development, with consequences that demand serious, evidence-based governance.

Readers who identify factual inaccuracies are encouraged to contact the author so that corrections may be incorporated into future editions.

PREFACE

My aunt has never owned a smartphone.

In 2015, she was in her fifties, living outside Filabusi in the dry cattle country of Matabeleland South, Zimbabwe and a member of a small cooperative of village women who grew vegetables and sold them at the local market. She carried a basic handset — the kind that makes calls, sends texts, and does very little else. It had a cracked plastic casing and a battery that needed coaxing. It was not, by any conventional measure, a financial instrument.

And yet, at the end of each market day that year, she would dial *151# and work through a short menu of numbered options — press 1 to send money, press 2 to check balance, press 3 to withdraw — and there, on a screen barely larger than a matchbox, she could see exactly where she stood. What the cooperative had earned. What she had saved. What her children in Harare and her niece in Johannesburg had sent directly to her account that week.

This was EcoCash. No bank branch, no forms, no minimum balance, no manager deciding whether she was creditworthy. Just a USSD string — a sequence of numbers punched into any handset on the Econet network — and a connection to a financial system that had, finally, been designed to reach her.

What the technology gave her wasn't sophistication. It was visibility. For the first time, she could watch her savings grow toward something specific — school fees, mostly, for the next generation. The women in the cooperative would check their balances between customers, comparing figures, encouraging each other. Remittances that had previously travelled through informal networks — with uncertain timing, unpredictable deductions, and no record of arrival — now landed directly, instantly, and legibly on that cracked little screen. A decade earlier, none of it had been available to her. The nearest bank branch was forty kilometers away over roads that flooded in the rains. The formal financial system required documentation she did not have, minimum balances that would have consumed a month of vegetable margins, and a physical presence in a building she couldn't easily reach. It had been designed, in every practical sense, for someone else.

EcoCash changed that — not with an app, not with a smartphone, but with twelve digits and a hash key.

This is the story that any honest account of fintech must begin with. Not with the billion-dollar valuations, the venture capital feeding frenzies, or the glossy product launches in San Francisco and London — though those are part of the story too. But with my aunt in Filabusi and the hundreds of millions of people like her, across Zimbabwe, Kenya, India, China, Brazil, and beyond, for whom the digital revolution in finance hasn't been a convenience upgrade but a categorical transformation.

The question this book asks is what comes next — and the answer is both more exciting and more dangerous than the fintech revolution's most enthusiastic advocates have acknowledged.

The first wave of fintech was about access. Getting the infrastructure — the mobile money rails, the digital wallets, the instant payment systems — to places the old system had never reached. That wave has been, by any honest measure, a remarkable success. More than a billion people have entered the formal financial system since 2011. Mobile money has become the primary financial institution for hundreds of millions of people across Sub-Saharan Africa and South Asia. Digital payments have compressed the cost and friction of financial transactions in ways that benefit not just individuals but entire economies. The evidence for these gains isn't anecdotal. It is rigorous, peer-reviewed, and accumulated across multiple geographies and methodologies.

But access, as the argument here is at some length, is only the beginning.

The second wave — the one we are living through now — is powered by artificial intelligence. And it is qualitatively different from what came before. The first wave extended the reach of financial services. The second wave extends their intelligence. It creates the possibility — still only a possibility, contingent on choices that humans are making right now — of a financial system that understands each of its users as individuals: their circumstances, their goals, their vulnerabilities, and their potential. An AI that can deliver personalized financial guidance in the right language, at the right moment, tuned to the specific situation of a first-generation bank customer in rural Zimbabwe or a gig economy worker in São Paulo — that AI isn't a luxury feature. It is potentially the most powerful tool for human economic empowerment that has ever been built.

The subtitle of this book is *Fintech's AI Inflection Point*. I chose that phrase deliberately. An inflection point is more than a moment of change. It's a moment at which the direction of change itself changes — where the curve bends, where what was true before becomes, if not false, then insufficient. The fintech story up to now has been, broadly, a story of expanding access and reducing cost. The AI inflection point changes that story into something simultaneously more ambitious and more treacherous: a story about intelligence, personalization, and the governance of systems that make high-stakes decisions at a speed and scale that no previous generation of financial technology has approached.

I have spent years studying this inflection point — in a large-scale empirical investigation of fintech's impact on financial development across 290 Chinese cities, and in a detailed analysis of what artificial intelligence in financial services actually does to the risk profile of the banks that must simultaneously compete with it, adapt to it, and be transformed by it. What I found isn't what the simplest versions of either the enthusiast or the sceptic account would predict. It is more complicated, more interesting, and — once understood — more useful for thinking about what should be done.

The short version is this. AI in financial services raises risk before it reduces it. The early phase of AI adoption — when banks are learning to use new tools, when credit models are immature, when the competitive pressure from fintech challengers is forcing decisions that prudence wouldn't — is measurably riskier than the phase that follows. But the phase that follows is clearly better: lower bank risk, better credit decisions, more accurate fraud detection, cheaper and more inclusive financial services. The journey matters enormously. So does the governance of the transit.

Any honest preface must also speak plainly about what could go wrong.

The same algorithmic intelligence that extends credit to the previously excluded can embed historical discrimination into automated decisions that process millions of cases per second with no human review. Platforms that use AI to personalize financial offers can use the same personalization to identify psychological vulnerabilities and target high-cost products at moments of greatest financial anxiety — something that the most sophisticated behavioral economists and product designers in Silicon Valley have become very good at. The concentration of AI capability in a small number of very large platforms creates system-level dependencies that regulators are only beginning to understand. The convergence of AI models across institutions creates a new and poorly appreciated form of correlated risk: when every bank's credit model is a variation of the same design, trained on similar data, the financial system loses the diversity that historically absorbed shocks rather than amplifying them.

These aren't hypothetical dangers. They are the documented findings of the most serious research institutions in global finance — the BIS, the IMF, the FSB — from people who are paid to worry about the stability of the financial system and who are, right now, really worried.

The present work is for anyone who wants to understand the inflection point we are at — what it means for the financial system, for the people that system is supposed to serve, and for the choices that will determine whether the AI-powered future of finance is an upgrade or a catastrophe. It is written to be read by the generalist who wants to understand why this technology matters, the practitioner who needs to manage it, and the policymaker who has to govern it. The academic detail is there for those who want it. So is the story.

Because the story — of my aunt in Filabusi pressing *151# on a cracked handset after market day, of the algorithm that extended credit to half a billion people in China, of the European bank that discovered its AI was raising its risk profile before it discovered that the same AI would eventually lower it — is, in the end, the defining thing. Finance, at its best, is the infrastructure of human possibility. The real issue is whether the AI inflection point extends that infrastructure or whether it rebuilds it in a form that serves only those who were already well served.

The tools are there. The puzzle is whether the will is.

A NOTE ON SOURCES AND METHODS

The present work draws on numerous sources of empirical research; however, core to the study are two of the author's original works. It starts with a study of fintech's impact on financial development across 290 Chinese cities and 31 provinces, covering the years 2011 to 2018. The methodology employs two-stage least squares instrumental variable regression, corrected for cross-sectional dependency using Pesaran's CD test and slope homogeneity tests. This study has been peer-reviewed and published in Financial Innovation. Next comes an analysis of fintech AI development and bank risk across 85 European banks in 21 countries from 2005 to 2022, using fixed-effects panel estimation and instrumental variable regression.

Where the findings of these studies are presented in the text, they are identified clearly. All other claims are supported by published academic research, official data from recognized international institutions, or primary sources from the organizations discussed. Case studies draw on publicly available information supplemented where possible by practitioner engagement.

The book is written to be read without specialist knowledge. Technical concepts are introduced where necessary and explained in plain language. Readers who want the full methodology are directed to the published papers listed in the notes.

A word on the subtitle: Fintech's AI Inflection Point is an original formulation of the argument central to this book — that we aren't living through a continuation of the fintech story but a qualitative change in its direction and implications. The phrase belongs to this work.

For my daughter, Amani Jordan,

who showed me how bright the stars shine in the African night sky

— and why their light belongs to everyone.

Table of Contents

CHAPTER ONE

From Fintechs to Techfins

When Giants Pivot

At its core, money is information. Increasingly, the institutions processing that information are no longer banks.

It was my first Chinese New Year. I had only been in China for four months, living in a relatively small city called Zhenjiang in Jiangsu Province. It was an unusually cold February, but the mood everywhere was warm and festive. For most of us, New Year means fireworks, resolutions, and promises to ourselves that we rarely keep. The fireworks were there, but something else caught my attention.

My WeChat app kept buzzing with messages from friends and colleagues. At first, I assumed they were the usual New Year greetings. But when I opened them, I realized they were something different — digital red envelopes. Each one contained a small, randomly allocated amount of money that appeared instantly in my WeChat wallet. For the first time, I had experienced a tradition that stretches back more than two thousand years: the sharing of money in red envelopes to celebrate the new year. Only this time, the envelopes were entirely digital.

In January 2014, WeChat introduced this feature. Known in Chinese as *Hong Bao*, it allowed users to send small amounts of money to each other — the digital equivalent of slipping cash into a birthday card. It was playful, almost trivial. A social feature designed for games between friends, for teenagers, for the exchange of tiny sums in a virtual world.

Nobody in a bank boardroom noticed. There was no reason they should have.

In 2014, the idea that a messaging application used by young people to send digital pocket money would eventually reshape parts of the global financial system was not a prediction anyone was making. It was not even a hypothesis. Yet just over a decade later, WeChat processes billions of financial transactions every day. The idle balances left in users' digital wallets have become the foundation for investment and lending products that extend credit to hundreds of millions of people the traditional banking system had never reached.

And the habit of reaching for a phone to pay for everything — from a bowl of noodles in a Beijing hutong to a property purchase in Shanghai — has become so deeply embedded in everyday life that cash itself has begun to feel like a relic.

This isn't a story about disruption in the way that business schools teach disruption. It isn't a story of a plucky challenger company attacking the market share of an entrenched incumbent. It's a story about something more fundamental and more troubling for the institutions that have governed money for centuries: the gradual, almost invisible colonization of financial services by companies that were never designed to be financial institutions at all.

The distinction matters enormously. It is, in fact, the central distinction that this book is built around.

There are two kinds of companies reshaping the global financial system — and they are doing so in fundamentally different ways.

The first kind — the startup challenger, the neobank founded in a co-working space, the payment processor built by two engineers with a better idea than Visa — starts with a financial problem. It sees that international transfers are too slow and expensive, or that small businesses can't get working capital, or that consumers deserve a credit card that doesn't exploit them, and it applies technology to solve that specific problem. This kind of company is, in fact, important. It has driven down costs, improved products, and forced incumbent institutions to confront how much friction and margin they had been extracting from their customers. But it is, ultimately, a financial company with better technology. It can be regulated like a financial company, absorbed by a financial company, or driven out of business by financial companies.

The second kind is different altogether, not just in degree. It starts with a technology platform that has already become central to daily life — a messaging service, a search engine, a commerce marketplace, a social network — and discovers, almost incidentally, that the platform it has built gives it the capability to provide financial services more cheaply, more conveniently, and more accurately than any bank has ever managed. This kind of company — the Techfin — doesn't disrupt financial services from outside. It occupies the layer of digital infrastructure that financial services increasingly depend on, and it does so from a position of competitive advantage that no amount of bank digital transformation can fully replicate.

Understanding this distinction is the key to understanding why the fintech era has produced not just new products but a new map of power in the global financial system — and why the arrival of artificial intelligence is about to redraw that map again.

THE ANATOMY OF A TECHFIN

$3Trillion+	5-6B	80%	6
Estimated size of the global fintech market by 2035	*Active digital wallet users worldwide by 2030*	*of Gen Z globally prefer digital-only banking*	*Big Tech firms now holding full banking licenses*

What makes a Techfin different from a fintech isn't primarily its technology. It is its starting position.

When Amazon began offering loans to marketplace sellers in 2011, it wasn't doing anything that any bank's small business lending team couldn't do in principle. Amazon could extend credit, yes. But what Amazon had that no bank possessed was a perfect, real-time, granular picture of every seller's business performance: their revenue, their inventory turns, their customer satisfaction scores, their fulfilment speed, their seasonal patterns, and their competitive position relative to every other seller in the same category. A bank extending a small business loan would ask for three years of financial statements and make an educated guess. Amazon already knew, to the day, whether that seller's business was growing, plateauing, or in trouble — and it knew this before the seller's accountant did.

This is what separates Techfin from fintech, and what makes the distinction so significant: the data advantage isn't just quantitative — more data than a bank has — but qualitative. It is data of a different kind. Behavioral rather than transactional. Predictive rather than descriptive. And it turns out that creditworthiness — the ability to repay a loan — is a behavioral phenomenon as much as a financial one. It shows up in how promptly you answer customer messages, in whether your inventory patterns suggest competent stock management, and in whether the reviews you receive indicate a business with loyal repeat customers or one surviving on volume alone. Banks weren't measuring these things because they could not. Amazon measured them continuously as a side effect of running its marketplace.

A significant number of studies show that the Techfin data advantage over incumbents is now substantial. Companies like Amazon have real-time data on the decisions of millions of small businesses, something which traditional banks do not necessarily have access to. This affects banks' accuracy in risk assessment compared to Techfins, which have richer behavioral data on their users. This translates into a three-fold advantage. Put simply, more accurate personalized data, more direct distribution, and more complete risk assessment.

THE DISTRIBUTION ADVANTAGE

The data advantage is only half of what makes a Techfin formidable. The other half is distribution.

Banks spend, on average, $300 acquiring a new customer in the United States — a figure that accounts for branch infrastructure, marketing, onboarding costs, and the compliance burden of establishing a new relationship. The economics of this acquisition cost create a structural bias toward customers whose lifetime value justifies the spend: higher-income customers, customers who will take multiple products, and customers who will stay for years. Someone who wants a basic savings account with a small balance is a poor investment of a $300 customer acquisition budget.

When WeChat launched its financial products in 2014, it had instantaneous access to 800 million active users at zero incremental acquisition cost. The financial service rode on the back of a messaging platform that users had already integrated into the fabric of their daily lives — one that they opened dozens of times a day, not to use financial services, but to talk to their family, pay for lunch, and watch short videos. The trust and habitual engagement that a bank would spend years and hundreds of dollars trying to build was already there, accumulated through years of social connection.

This distribution advantage isn't replicable by any amount of marketing budget or product redesign. You can't buy the daily habitual engagement that comes from being the primary communication medium between people and the people they love. The Techfin platform doesn't compete for customer attention. It already has it and has had it long enough that the transition to financial services feels natural rather than intrusive.

What follows from this — and this is the insight that bank strategists took too long to absorb — is that the competitive threat from Techfin isn't primarily about products. In most cases, Techfin products are not materially superior to bank products. WeChat Pay isn't a better current account than HSBC's. Ant Group's loans aren't structured differently from a bank loan. What is different is the context in which those products are encountered, chosen, and used. When payment is the natural continuation of a conversation, when investment is something you do with the balance sitting in your messaging wallet, when insurance appears at the moment you are booking the holiday you would need it for — the act of choosing a financial product has been so embedded in the flow of daily life that it barely registers as a choice at all.

This is what banks are competing against. Not a better product. A better context.

CASE STUDY — INDIA: THE PUBLIC INFRASTRUCTURE ANSWER

India's response to the Techfin challenge offers a model that is, in some ways, weightier than either the Chinese or American version of the story — because it demonstrates what becomes possible when the infrastructure advantages that Techfins exploit privately are instead built as public goods.

The India Stack is a layered design of digital public infrastructure, assembled over roughly a decade from 2009 to the present. Its foundation is Aadhaar, a biometric identity system that has enrolled more than 1.3 billion Indian residents — the largest national identity program ever undertaken. Built on that foundation is the Unified Payments Interface, a real-time payment system that any registered bank or payments company can connect to as a shared utility. Above that sits DigiLocker, a cloud-based document storage system that allows citizens to hold and share official documents digitally. And above that, the most recent addition is the Account Aggregator framework — a consent-based data sharing design that allows individuals to authorize any financial institution to access their financial history across all of their accounts with a single digital permission.

The aggregate effect of this structure is extraordinary. A person who has no formal credit history but has been transacting through UPI for three years has, in those transaction records, something more informative than any traditional credit file — because the file reflects actual behavior rather than the self-reported summary that a traditional credit application requires. A business that has been filing GST returns for two years and collecting payments through UPI has, in those records, a fuller picture of its financial performance than any three years of bank statements could provide. And any authorized lender — fintech or bank, established or startup — can access that picture with the customer's one-click consent, in seconds, without a single paper document.

What follows is that the data advantage that Ant Group spent a decade building privately, and that Amazon built as a side effect of running its marketplace, is available in India to any participant in the financial system, on equal terms. The playing field isn't perfectly level — large platforms still have the distribution advantages that come from operating the interfaces through which people interact with the Stack — but the informational monopoly that allows a WeChat or an Ant Group to set terms for the entire financial system is significantly reduced.

UPI processed over 200 billion transactions in the twelve months of 2025. It now accounts for 84% of all digital payments in India and more than 46% of all real-time payment transactions globally. PhonePe, Google Pay, and Paytm collectively serve more than 450 million active users — but they do so on shared rails that the central bank built and owns, not on proprietary infrastructure that locks users in.

The India model is the most significant policy response yet developed to the Techfin challenge. It won't be the last.

WHY GOOGLE DOES NOT NEED TO BE A BANK

Among the technology companies that have entered financial services, Alphabet's approach is the most strategically sophisticated — and in some ways the most significant — exactly because it requires Google to be nothing so obvious as a bank.

Google controls the search interface through which consumers discover financial products. It controls the advertising system through which every bank and insurance company markets to those consumers. It provides the cloud infrastructure on which an increasing share of the financial system's operations run. It processes more than 200 billion searches per month, a significant proportion of which are directly financial in nature — queries about mortgage rates, credit card comparisons, savings account options, insurance quotes. And through Android and Google Pay, it has established a presence in the payment layer that handles hundreds of millions of transactions daily.

Google doesn't need to hold deposits or extend credit to capture a substantial share of the value that the financial system generates. It captures that value upstream — in the advertising revenues of every financial institution competing for customers through Google's platforms, in the cloud service fees of every bank that has migrated its infrastructure to Google Cloud, and in the data generated by every financial search and transaction that flows through Google's products.

This is why the competitive question banks should be asking about Google isn't "will Google launch a bank?" The more important question is "are we already paying Google rent for access to our own customers, and is that rent increasing?" The answer, in most sophisticated banking markets, is yes — and the proportion of bank revenue flowing to technology intermediaries is rising.

The Bank for International Settlements documented this structural shift in a 2022 working paper. Their core finding: as financial services migrate to digital channels, the customer relationship is increasingly owned by the platform that controls the interface, rather than by the institution that provides the underlying financial product. Banks become manufacturers. Platforms become distributors. The distribution margin — historically captured by banks through branch networks and relationship managers — transfers to platform owners. This isn't a future threat; it's the present reality, and it is accelerating.

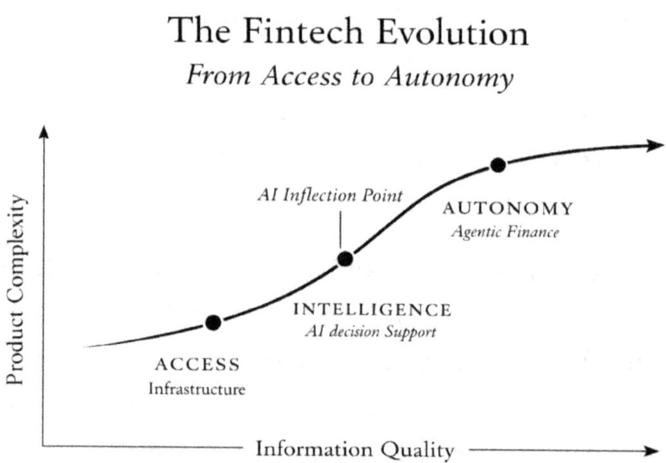

The Fintech Evolution
From Access to Autonomy

THE THREE REGULATORY PHILOSOPHIES

Financial regulators globally have watched the rise of Techfin with a combination of genuine admiration — these platforms have achieved inclusion goals that decades of official programs failed to meet — and genuine alarm, because the same platforms now occupy positions of system-level importance in financial systems around the world without being subject to the prudential regimes that apply to institutions of comparable importance.

Three distinct regulatory philosophies have emerged from the divergent national responses to this challenge and understanding them is essential context for everything that follows in this book.

The Chinese model is state-led and structurally decisive. When Ant Group's rapid growth was judged to have created system-level importance without commensurate accountability, the response was swift and extensive: the IPO was blocked, capital requirements equivalent to those of a bank were imposed, and data sharing between Ant's financial operations and Alibaba's commerce platform was restricted. The signal sent was unambiguous — no private platform, regardless of its apparent social utility, would be permitted to operate at system-level scale outside the prudential framework that the state applies to system-level important institutions. The costs were real: innovation slowed, Ant's growth was curtailed, and Jack Ma, one of China's most prominent entrepreneurs, effectively disappeared from public life for two years. But the system-wide risk posed by unregulated growth was addressed in a way that no other major jurisdiction has matched in both speed and scope.

The European model is rights-based and architecturally ambitious. The EU's regulatory response to digital finance is the most legally extensive in the world: MiCA for crypto assets, DORA for operational resilience, the AI Act for algorithmic systems, PSD3 for payments, and the Digital Markets Act for platform competition.

These regulations collectively constitute an attempt to build a coherent governance structure for the digital financial system before the risks crystallize rather than after — an approach that is historically unusual and requires a regulatory confidence in anticipating consequences that is difficult to sustain.

The American model is market-led and institutionally fragmented. The United States remains the most permissive major jurisdiction for fintech, partly because its divided regulatory structure — between the Fed, the OCC, the CFPB, state banking regulators, the SEC, the CFTC, and others — makes sweeping regulatory responses slow to develop and easy to fragment. The effect is the world's most vibrant fintech sector and the widest gaps in structural oversight. The $32 billion collapse of FTX, the largest single disaster in the history of digital assets, happened almost entirely within the gaps of this system.

None of these philosophies is obviously correct in all dimensions. The Chinese approach prioritizes stability and state control, sometimes at the cost of innovation and entrepreneurial autonomy. The European approach prioritizes rights and market fairness at the cost of regulatory complexity and compliance burden. The American approach prioritizes innovation at the cost of consumer protection and structural resilience. The regulatory challenge of the AI inflection point — which this book develops through the chapters that follow — is to find the path through this choice that captures the benefits of each without inheriting the worst of any.

That is harder than it sounds. But it's the work.

Ant Group: From Escrow Service to Financial Ecosystem

Alipay launched in 2004 as a simple escrow service — a trust mechanism for Alibaba's marketplace that held payment until buyers confirmed receipt of goods. Nobody planned a financial empire. The empire followed from the logic of the platform: once users trusted Alipay with their money, holding a balance there was natural. Once balances accumulated, investment products could be offered. Once investment was established, credit followed. Once credit existed, insurance made sense.

Yu'e Bao, the money market fund Ant launched in 2013, accumulated approximately 500 billion yuan (around $80 billion) in assets within nine months. By 2017, it was the world's largest money market fund — built not by marketing to investors but by making it trivially easy for Alipay users to earn interest on money they were already keeping in their wallets. The minimum investment was one yuan. The minimum return period was one day.

Ant's micro-lending platform extended credit to hundreds of millions of individuals by 2023 — more than the combined consumer lending customer base of the five largest US banks. The key to this scale isn't capital efficiency but information efficiency. Ant's credit models, trained on transaction data, social data, and behavioral data from across Alibaba's market, can assess a street vendor's creditworthiness with greater accuracy than any bank loan officer, using information the bank could never collect.

The Ant Group restructuring of 2020–2023 — which saw Chinese regulators impose bank-equivalent capital requirements and restrict data sharing between Ant's financial operations and Alibaba's commerce platform — established a precedent with global significance: that the structural importance of a Techfin platform, regardless of its legal structure, can justify the full weight of financial regulation.

TAKEAWAY ANT: *Ant Group's twenty-year arc from escrow service to restructured financial holding company illustrates both the extraordinary velocity of Techfin growth and the inevitability of regulatory reckoning when that growth becomes structurally significant.*

India Stack — Aadhaar biometric identity, UPI payments, DigiLocker, and the Account Aggregator framework — is one of the most significant public digital infrastructure investments of the decade. UPI processed 131 billion transactions in the twelve months to March 2024, representing over 46% of all global real-time payment transactions.

The Account Aggregator framework, launched in 2021, provides consent-based data sharing: a business owner with three years of GST filing history can establish creditworthiness with a lender in minutes, without producing a single paper document. The information was always there. The infrastructure to share it securely and consensually was not.

The India model answers the question that the Chinese and American models cannot: what happens to the development dividend when platform companies, rather than the public, own the digital infrastructure of financial services? India's answer is that public infrastructure, designed with open access principles, can capture the efficiency gains of digital finance while distributing its benefits more equitably than any private platform is incentivized to do.

TAKEAWAY India *Stack demonstrates that the most powerful fintech infrastructure is public rather than proprietary — designed to multiply the productivity of every application built on it, rather than to lock users into a single platform's services.*

CHAPTER TWO

Fintech and Financial Development

Evidence from China and Beyond

The chief lesson from two decades of fintech is that financial development isn't just a financial problem. It's an information problem — and fintech has become the most powerful information infrastructure ever built for financial markets.

900M+	50%+	89%	$2Trillion+
Chinese consumers actively use mobile payment platforms	*of global digital payment transaction value now flows through China*	*of Chinese adults banked by 2021, up from 35% in 2010*	*in annual digital loans have been issued by Chinese fintech platforms*

The most extraordinary natural experiment in the history of financial development began on 13 June 2013, when a company called Tianhong Asset Management launched a money market fund called Yu'e Bao. The name, in Mandarin, translates roughly as Leftover Treasure. The name was deliberately unpretentious because the product was designed for unpretentious people: Alipay users who kept small balances in their digital wallets — balances they were neither saving intentionally nor spending immediately, money that in the old financial world would simply have sat idle, earning nothing.

Yu'e Bao took those idle balances and put them to work, paying a daily interest rate that, at the time of launch, was approximately three times what the major Chinese state banks paid on a one-year fixed deposit. The minimum investment was one yuan — about fourteen cents. There were no fees, no minimums, no lock-up periods. Money could be added or withdrawn in amounts as small as one yuan at any time.

Within a year, Yu'e Bao had over 100 million customers and 92 billion yuan in assets under management. By 2017, it was the largest money market fund in the world, with more assets than the bond funds of several European countries. The people who built it hadn't intended to create the world's largest money market fund. They had intended to make Alipay more useful by giving idle balances a return. The scale emerged from the logic of the platform: once the product existed, every idle balance in every Alipay wallet was a potential investment, and there were three hundred million Alipay wallets.

This is the story that makes financial economists put down their coffee and stare at the wall. Not because the product was sophisticated — it was not. Money market funds have existed for decades. But because the scale at which the platform made it possible to reach customers who had never previously participated in investment markets — never held a financial product more complex than a bank account, never considered themselves investors — was unlike anything in the prior history of financial inclusion.

And Yu'e Bao wasn't even the main event.

THE EVIDENCE FROM CHINA

The author's own research examines the full picture — not just the investment products, but the entire landscape of fintech's impact on financial development across 290 Chinese prefecture-level cities over eight years. The scope reflects the ambition of the question: not "did this specific product work?" but "what did the systematic penetration of digital financial services do to the financial development of one of the world's largest and most diverse economies?"

The methodology requires some explanation, because it determines the true meaning of the results. The central challenge in any study of fintech's effects is distinguishing the impact of fintech from the impact of everything else that was happening simultaneously — economic growth, urbanization, education expansion, and dozens of other forces that both drive fintech adoption and independently affect financial development. To isolate the effect of fintech, the study uses instrumental variable regression: it identifies factors that drove the spread of fintech adoption in some cities but not others for reasons unrelated to those cities' financial development trajectories — specifically, early mobile phone and internet penetration, which gave some cities a head start in the digital infrastructure required for fintech to operate.

The logic is this: a city that happened to have better mobile connectivity in 2008 — perhaps because it was on a major transport corridor or a university town with research infrastructure — adopted fintech earlier and more extensively than comparable cities without that head start. That earlier adoption wasn't caused by the city being already more financially developed. It was caused by an infrastructure advantage that, for the purposes of the study, was essentially random with respect to financial outcomes. By using this early-connectivity instrument, it is possible to identify the causal effect of fintech penetration on financial development, rather than merely observing their correlation.

The primary finding is unambiguous and robust to multiple specifications: fintech development has a significant, positive, and causal effect on financial development outcomes. Cities that experienced higher fintech penetration showed stronger growth in loan access, deeper deposit mobilization, wider availability of insurance and investment products, and faster expansion of the financial sector as a whole. This relationship holds after controlling for income, population, urbanization, and province-level fixed effects. It holds when the sample is restricted to less-developed cities. It holds when the analysis is conducted separately for the fintech components of breadth, depth, and digitization.

THE THREE-DIMENSIONAL FINDING

Here is what the data actually shows.

Fintech development, when measured carefully and its causal effect isolated from everything else happening simultaneously in a rapidly changing economy, does more than extend the reach of the financial system. It deepens it. The China study's primary result isn't a single finding. It is three findings that arrive together, reinforce each other, and together say something more important than any one of them could say alone.

The first finding is about credit. In cities where fintech penetration grew faster — controlling for income, urbanization, province-level policy, and dozens of other variables — loan balances at financial institutions grew faster too. A 10% increase in fintech development is associated with a 7.5% increase in financial access. That isn't a marginal effect. It's the difference between a financial sector that reaches most of its potential borrowers and one that reaches a fraction of them. The mechanism is the one fintech's advocates have always described: digital platforms slash the cost of credit assessment, documentation, and disbursement. What changes in the data is that the mechanism can now be shown to work, at scale, over time, and in the presence of rigorous controls. The argument is no longer theoretical.

The second finding is about deposits. The same fintech penetration that expands credit supply also draws more household savings into the formal financial system. In cities with greater fintech development, deposit balances at financial institutions grew more quickly. This matters in ways the access narrative tends to underemphasize. Deposits aren't simply a funding source for banks. They are evidence of a population that has chosen to store its wealth formally rather than informally — in accounts rather than under mattresses, through institutions rather than through rotating credit groups or livestock or land. That choice, when it happens at scale, changes the texture of an economy. It deepens the resource base from which credit is extended. It connects households to a financial system that can serve them beyond their immediate transaction needs. And it did not require any policy initiative other than giving people access to digital financial tools they found useful enough to trust.

The third finding is the one that receives the least attention and deserves the most. In cities where fintech penetrated more deeply, household savings rates — the share of income that households set aside as a buffer against future uncertainty — were higher. Fintech isn't just widening access to credit. It is building financial resilience. It is making households less vulnerable to the kind of income shock — a medical bill, a failed crop, a month without work — that can undo years of hard-won economic progress in a single blow. The effect size is almost identical to the deposit effect. It isn't an afterthought in the data. It's a co-equal result.

Put the three findings together and you have something that looks less like a fintech story and more like a financial development story. Access, depth, and stability: these aren't three different products from the same platform. These are three dimensions of what a functioning financial system is supposed to do for the people it serves. The China study finds that fintech advances all three, simultaneously, causally, and at a scale that is hard to dismiss. That is the central result. Everything else in the chapter builds on it.

It is worth dwelling on why the three-part structure matters. A study that found fintech only improved credit access would be good news for borrowers and for banks willing to serve them, but it would leave open an uncomfortable question: are we building a financial system that extends credit to people without building the savings culture and resilience that make that credit sustainable? A study that found fintech only deepened deposits without expanding access would suggest a tool that mostly serves those who are already financially engaged. The China data doesn't leave either question open. It finds that fintech does all three things at once — and in doing so, it resolves what had been one of the field's most persistent theoretical anxieties. Fintech isn't trading stability for access. It is delivering both, alongside something that looks very much like security.

This isn't the story that was being told a decade ago, when the financial inclusion debate was organized almost entirely around a single metric: the percentage of adults with a formal bank account. That metric is important. But it measures presence, not engagement. It counts the door without asking whether anyone walked through it, and if they did, whether there was anything worth finding on the other side. The Chinese data counts what happened inside.

THE REGTECH FINDING

The second major finding from the China study deserves separate treatment because it bears directly on the regulatory question that consumes much of the second half of this book.

The study finds that regulatory technology — the application of technology to the regulatory and compliance infrastructure of the financial system — is an independent driver of financial development outcomes, above and beyond the direct effect of fintech itself. Cities in provinces with better-resourced financial regulatory bureaus, measured by the number of regulatory staff per financial institution and by the sophistication of supervisory reporting systems, experienced larger fintech-development dividends than comparable cities with weaker regulatory infrastructure.

This is counterintuitive from the perspective of the standard regulatory narrative, in which regulation is a cost imposed on a productive system. The Chinese data suggests a different relationship: better governance of the fintech sector produces more financial development because better-governed products are more trusted, more actively used, and better designed. When consumers know that a product is subject to meaningful oversight, they engage with it more confidently. When providers know that their products will be evaluated against conduct standards, they design them more carefully. The invisible hand needs a visible referee.

This finding — that regulatory quality is a complement to fintech investment rather than a substitute for or constraint on it — is one of the key insights in this book. It will recur in different forms in the chapters on AI risk, on system-wide risk, and on the regulatory response.

FROM CHINA TO KENYA: THE M-PESA COMPARISON

No accounting of the evidence on fintech and financial development would be complete without a thorough treatment of M-Pesa, the Kenyan mobile money platform that, in the literature, is one of the most studied examples of digital financial inclusion in the world. The scale of M-Pesa's adoption — from zero to 90% of Kenyan adults in roughly twelve years — and the quality of the academic research it has generated make it the indispensable reference point.

The foundational research by Tavneet Suri and William Jack, published in *Science* in 2016, is by now well enough known that its headline numbers have become a staple of fintech presentations worldwide: 2% of Kenyan households lifted out of extreme poverty, a roughly 6% average consumption increase for households with M-Pesa access, and an 18.5% consumption increase for female-headed households. But the headline numbers obscure several features of the research that matter enormously for how its conclusions should be interpreted.

First, the consumption gains flow primarily through the insurance-like function of mobile money — the ability to receive small transfers from the social network during economic shock — rather than through the savings or credit products that M-Pesa subsequently developed. The poverty reduction documented by Suri and Jack is poverty reduction through resilience: the ability to absorb a medical bill, a bad harvest, or a job loss without a catastrophic reduction in living standards. Here is a specific and important kind of financial development, but it is different from the credit-driven business investment that financial inclusion programs sometimes present as the primary pathway from poverty.

Secondly, the female household effect — the approximately 18.5% consumption gain for female-headed households versus the roughly 6% average — is not primarily about money market funds or investment products. It is about financial autonomy. When a woman can receive payment, save, and deploy money through a digital account that belongs to her alone — one that requires no co-signer, no male relative's knowledge, and no physical branch visit — something changes in the economic structure of her household. The channels run through consumption choices, investment in children's education and health, and the economic resilience that comes from having a private resource that can't be appropriated. This isn't primarily a story about financial technology. It is a story about financial autonomy — and it's a story that AI-powered financial services could tell at far greater scale than M-Pesa alone.

Third, the research demonstrates that the interaction of digital money and agent density networks was one of the main causal variables, not necessarily app downloads. The study's identification strategy uses the spread of mobile money agents as a source of variation, not account sign-ups, nor transaction volumes. Suri and Jack found that areas with higher agent densities saw greater poverty reduction. This says a lot about the network's physical infrastructure, which can be summed up as human beings in a kiosk. In other words, what drove outcomes was not necessarily the digital technology in itself, but its interaction with reliable access to an agent cash-in and cash-out network.

THE INDIA STORY: UPI AND THE FINANCIAL INCLUSION SPEED RUN

If Kenya's M-Pesa demonstrated that digital financial infrastructure could produce genuine development impact in Sub-Saharan Africa, India's UPI has demonstrated that the same logic, applied through public infrastructure rather than a private platform, can operate at a scale that M-Pesa never approached.

The Unified Payments Interface reached its first billion transactions in 2017. By 2024, it was processing more than 14 billion transactions monthly. The Jan Dhan, Aadhaar, Mobile trinity — the Indian government's description of the three pillars of its digital inclusion strategy — has opened more than 520 million bank accounts since 2014, connected hundreds of millions of account holders to the Aadhaar biometric identity system, and given each of them access to the UPI payment network through whatever mobile device they own.

The economic consequences are beginning to show in the data. Research by Ratna Sahay and colleagues at the IMF, examining the relationship between financial inclusion and economic growth across developing economies, finds that financial inclusion boosts growth, but with diminishing returns. There are significant positive effects of financial inclusion on economic growth across a broad cross-country sample. But the relationship isn't linear — the benefits taper off as inclusion and financial depth increase. This is the "too much finance" finding, applied specifically to inclusion. It means the biggest growth payoff comes from moving people from zero access to basic access, not from deepening services for those who are already included. The marginal return on inclusion seems to be highest in countries that are furthest behind. This simply means piling on more financial services beyond a threshold doesn't keep delivering good results.

A good example is credit inclusion. Where supervision is weak, increasing access to credit can compromise economic and bank stability. The quality of bank supervision determines the resulting cost of expanding financial inclusion. Put simply, countries that have better bank supervision suffer the least from the negative impacts of deepening financial services. The opposite is true for low quality supervision: an increase in financial inclusion would inevitably result in greater instability. It's not that inclusion causes instability; it's that inclusion without governance causes instability. A trade-off we will explore later in the book.

It is also important to note that this trade-off depends on the form of inclusion. Non-credit forms of inclusion, such as access to and use of bank accounts, branches, and ATMs — do not necessarily hurt stability and can be promoted extensively. This is the finding that rarely gets quoted but changes the policy calculus entirely. Savings accounts, payment services, and transaction infrastructure carry no measurable stability cost. The risk sits specifically in credit expansion, not in inclusion broadly. For policymakers, it means you can push hard on payments, savings, and insurance without worrying too much about the stability trade-off — but the moment you extend credit at scale, supervisory quality becomes the binding constraint.

The results are still accumulating. But the directional evidence is strong enough to support a conclusion: fintech, when deployed thoughtfully on well-designed infrastructure with appropriate regulatory governance, does what its most enthusiastic advocates have claimed for two decades. It extends the circle of financial participation. It deepens economic resilience. And it does so most powerfully for the populations whose exclusion from the financial system has been most severe.

The question that follows — the question this book's remaining chapters pursue — is what happens when artificial intelligence enters this story. Because the AI inflection point changes the terms of the analysis in ways that the access-centric narrative of the first fintech era hasn't fully absorbed.

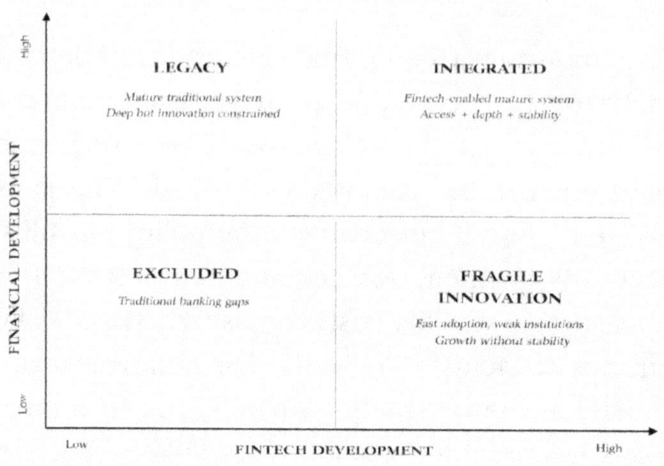

Research Finding: Fintech and the Three Dimensions of Financial Development

Muganyi et al. (2022) studied fintech's effect on financial development across 290 Chinese cities from 2011 to 2018. To deal with the endogeneity problem — the possibility that already-developing cities simply adopted fintech faster, rather than fintech causing their development — the study uses two-stage least squares with early mobile infrastructure as an instrument. The results are clear and consistent across all three outcomes.

Financial access (loan balances). A ten percent increase in fintech development raises loan access by 7.5%. This is the largest single-dimension effect in the study and confirms, in a rigorous causal framework, that fintech actually expands credit supply rather than merely relocating it.

Financial depth (deposit balances). The same fintech penetration that widens credit access also deepens the deposit base. The coefficient is 0.45 — for every 10% rise in fintech development, deposit balances grow by 4.5%. A deeper deposit base is more than an accounting outcome: it represents households actively choosing to store value within the formal financial system.

Financial stability (household savings rates). A coefficient of 0.44 on the stability dimension is statistically indistinguishable from the depth effect and almost as large as the access effect. Fintech is building financial buffers, not just financial products.

The study also finds that regulatory technology — the application of fintech tools to financial oversight — independently boosts all three dimensions. Its largest effect is on credit access (coefficient 0.455), and its presence in the model suggests that the quality of financial governance isn't a brake on fintech development but a driver of it. Better-governed markets generate larger fintech dividends.

The policy implication follows directly. Fintech investment that's not matched by investment in regulatory quality and institutional capacity will capture only a fraction of the available gains. The full dividend — expanded access, deeper deposits, greater resilience — accrues to markets that build all three foundations simultaneously.

Source: Muganyi, T., Yan, L., Yin, Y., Sun, H., Gong, X., & Taghizadeh-Hesary, F. (2022). Fintech, regtech, and financial development: evidence from China. Financial Innovation, 8(29).

CHAPTER THREE

Transforming Incumbents

The Disruption from Within

The most dangerous competitor isn't the one attacking from outside. It's the one who has made you irrelevant before you noticed the fight had begun.

$385B+	70%	5000+	45%
Global bank technology spending, 2026	of major banks with dedicated fintech innovation units	Active bank-fintech partnerships globally	Cost reduction achievable through full digital transformation (McKinsey)

In April 2018, a bank called TSB attempted one of the largest technology migrations in British banking history. It was moving the accounts of 5.2 million customers from a system built by its former parent company, Lloyds Banking Group, to a new platform built by its current parent, Banco Sabadell of Spain. The migration had been planned for years, tested extensively, and was considered by those responsible for it to be well-prepared.

Over the weekend of 20 April 2018, TSB threw the switch.

What followed was one of the most visible and costly technology failures in the history of British finance. Customers attempting to access their accounts found themselves locked out. Some could see their own accounts but also the accounts — and, in some cases, the credit card numbers and transaction histories — of strangers. Others attempted to transfer money to their own accounts and found the transfers disappearing. The bank's mobile app failed. The telephone helplines collapsed under the volume of desperate calls. In some documented cases, customers found their overdraft limits had vanished, leaving them unable to meet standing orders for rent, utilities, or loan repayments.

The crisis lasted not for hours but for weeks. By the time TSB's systems were functioning normally, the bank had lost more than £330 million — a figure that accounted for direct remediation costs, lost business, regulatory fines, and the incalculable long-term damage to customer trust. The bank's CEO resigned. The regulators commenced enforcement actions. And the TSB migration entered the curriculum of every bank technology program in the world as the definitive example of how not to modernize a financial institution's infrastructure.

But the lesson most people took from TSB was the wrong one.

The lesson most organizations drew was technology migrations are dangerous, so proceed slowly and cautiously. This conclusion is not wrong. But it is treacherously incomplete. The more important lesson — the one that TSB's own post-mortem and the subsequent investigations documented — is that the danger in technology transformation isn't the transformation itself. It's the gap between the ambition of the transformation and the organizational readiness to execute it. TSB's leadership was committed to a new system. The technical team executing the migration was under-resourced and under-supported. The testing regime was inadequate for the complexity of the system being replaced. And the contingency planning — the plans for what to do when things went wrong — was insufficient for the scale of what actually went wrong. The migration did not fail because it was attempted. It failed because it was attempted badly.

What makes this important is the context in which it happened. TSB's migration disaster occurred in 2018, at a moment when the competitive pressure from fintech on traditional banking was becoming impossible to ignore. Digital challengers weren't simply offering better products. They were at root changing consumer expectations about what banking should feel like, how fast it should work, and what it should cost. TSB's attempt to modernize wasn't a choice made in a vacuum. It was a response to a competitive environment in which the cost of standing still was becoming as dangerous as the risk of moving.

This is the frame through which incumbent bank transformation must be understood. Not as a choice between the risk of change and the safety of stability — but as a choice between two different kinds of risk. The risk of transforming badly and the risk of not transforming at all.

THE FIVE FRONTS

Modern banking transformation isn't a single project. It's a simultaneous war on five fronts, each of which requires a different strategy, different capabilities, and different leadership. The banks that have managed it most successfully are those that understood this — that treated transformation not as a technology project with a completion date, but as a permanent orientation of the organization toward a future in which digital capability and financial credibility must coexist.

The first front is the legacy technology problem. At the core of every major bank lies an infrastructure so old that its replacement has been deferred for decades. COBOL code written by programmers now retired. Batch processing systems were designed when overnight reconciliation was the best available option. Interface structures built before the API was invented, before the smartphone was imagined, before the idea of a customer checking their balance in real time was anything other than science fiction.

The estimate most commonly cited by bank technologists is that the major global banks' IT budget comprises of 60% to 75% in accumulated technical debt — the cost of modernization deferred. This isn't a precise figure, though some estimates put it between $1.5 trillion and $2 trillion U.S. dollars at the time of writing. It's an order-of-magnitude estimate of a real problem: that the structure of global banking is built on foundations that weren't designed for the world in which banking now operates, and that rebuilding those foundations while the bank continues to serve millions of customers is among the most complex engineering problems in the world.

The banks that have done this most successfully — DBS in Singapore, ING in the Netherlands, BBVA in Spain — share a common approach. They did not attempt to replace their legacy systems wholesale. They built cloud-native capabilities alongside the legacy infrastructure, migrating products and customer journeys incrementally while maintaining continuity throughout the transition. They accepted that the migration would take years, not quarters, and that the goal wasn't to reach a completion date but to reach a state in which the legacy footprint was progressively smaller and the cloud-native footprint progressively larger.

This approach requires patience, sustained investment, and — most importantly — a board and executive team that understands that the return on this investment is measured not in the next quarter's earnings but by the bank's competitive position in five and ten years.

The second front is the open banking revolution. Beginning with the UK's Competition and Markets Authority in 2016 and extending through the EU's PSD2, Australia's Consumer Data Right, and now regulatory initiatives in more than 80 countries, the open banking movement has established a new principle: customers' financial data belongs to them, not to their bank, and they have the right to share it with any authorized party of their choosing.

The immediate consequence of open banking is the reduction of the information monopoly that has historically been one of banking's most powerful barriers to entry. A bank that has held your transactions for a decade knows things about your financial life that no competitor could replicate. Open banking — by mandating that this data can be shared with your permission — makes that knowledge portable. Your credit history can be used by a new lender who has never dealt with you before. Your spending patterns can feed an aggregation app that gives you a complete view of your finances across multiple institutions. Your payment history can serve as the basis for a credit assessment by a fintech lender with better models, even if there is no prior relationship with you.

Banks initially treated open banking as a regulatory burden — an obligation to share an asset they would rather keep. The smarter institutions have come to see it differently. By investing in open banking infrastructure and API quality, they have positioned themselves as the preferred data source in a market of third-party financial services — capturing revenue from providing data and services to the market rather than merely defending the products they hold.

In the UK, where open banking has operated longest, the pattern is becoming clear. More than 60 million consumers are now using open banking-enabled products, from account aggregation apps to credit decisions based on their actual transaction history rather than a static credit file. The banks that invested in high-quality API infrastructure — Lloyds, Barclays, and particularly Starling — have become significant participants in the market rather than merely its unwilling contributors.

CASE STUDY — DBS BANK: THE MAKING OF THE WORLD'S BEST DIGITAL BANK

In 2009, when Piyush Gupta became CEO of DBS, the Development Bank of Singapore, he took the helm of an institution that, by most assessments, was merely adequate. DBS was safe, reliable, and deeply boring — a bank that Singapore's middle class used because it was there, not because it offered anything they particularly valued.

What Gupta understood, and what shaped the transformation he launched in 2014, was that 'adequate' was a deteriorating condition rather than a stable one. The direction of digital expectations in Singapore's connected consumer market meant that a bank that stayed merely adequate would become progressively inadequate. The competitive pressure wasn't immediate — digital challengers hadn't yet mounted the assault on Southeast Asian banking that would come in the following decade — but it was visible on the horizon, and Gupta chose to run toward it rather than wait for it to arrive.

The transformation program he launched under the deliberately provocative banner of 'Making Banking Joyful' was built on four foundations that together constitute what DBS calls its GANDALF strategy — a comparison to the world's most powerful technology companies (Google, Apple, Netflix, DBS itself, Amazon, LinkedIn, Facebook) that would have seemed hubristic from almost any other bank CEO.

The first foundation was technology. DBS committed to rebuilding its core banking infrastructure on cloud-native principles — not the partial cloud migration that most banks pursued, in which some functions move to the cloud while core banking remains on legacy systems, but a full re-structure that placed cloud at the center of the institution's operating model. This involved replacing or retiring more than 1,600 legacy applications over five years, moving the majority of DBS's computing workload to cloud infrastructure, and building the API connectivity that allowed the bank to integrate its products and data with partners and platforms.

The second foundation was data. DBS built what it calls its 'data flywheel:' a system for collecting, cleaning, and deploying customer data that allows the bank to move from responding to customer requests to anticipating customer needs. When a DBS customer's spending patterns change in a way that historically predicts a mortgage enquiry — when they start researching property websites, when their savings rate increases, when they pay off a credit card balance — the bank can initiate a conversation about home financing before the customer has articulated the intention. When a business customer's cash flow patterns suggest a liquidity gap coming in the next 60 days, DBS can offer a working capital solution before the gap becomes a crisis.

The third foundation was culture. This was, by Gupta's account, the hardest change to make and the one that determined whether the technology investment would translate into a genuine competitive advantage. DBS introduced 'Kill DBS' workshops — facilitated exercises in which teams were asked to design a fintech startup that would put DBS out of business. The intent wasn't morbid. It was diagnostic: by building the strongest possible version of the disruption case, DBS could identify its most significant vulnerabilities and address them before a competitor did. The exercise also served a cultural function: it normalized the idea that the bank's current competitive position wasn't permanent, and that the enemy of transformation wasn't the regulator or the market — it was internal complacency.

The fourth foundation was market. Rather than attempting to build every financial product its customers might want, DBS built an API platform that allowed more than 500 partner organizations — from insurance companies to property platforms to travel businesses — to embed DBS financial products within their own customer journeys. A property platform can offer a DBS when a customer expresses purchase intent. A travel booking service can offer DBS travel insurance at checkout. A retail platform can offer DBS credit at the point of sale. The products are DBS's. The distribution is allocated among the partners. The value is captured by both.

By 2016, DBS had been named the world's best digital bank by Euromoney, Global Finance, and the Banker — simultaneously, a distinction no other institution in the world had achieved. Its cost-to-income ratio had fallen from 47% to 40%. Its return on equity had risen from 10% to 18%. And its customer satisfaction scores had transformed from below the Singapore banking average to consistently above it.

The DBS story is not a story about technology. Technology, as Gupta has said repeatedly, is the easy part. The hard part is the organization — the leadership, the culture, and the willingness to be honest about what needs to change and what must not.

THE NEOBANK CHALLENGE: WHAT CHALLENGER BANKS ACTUALLY CHANGED

The neobank movement — digital-only banks built from scratch for a mobile-first world — has been celebrated, criticized, misunderstood, and underestimated in roughly equal measure.

The truth about its impact is more specific and more interesting than either the enthusiasts or the sceptics have articulated.

Neobanks did not take significant market share from incumbent banks. Monzo, Revolut, Starling, and other EU Neobanks collectively serve more than 100 million customers across Europe — a number that sounds large until you consider that the four largest UK banks alone serve the vast majority of the country's 70 million residents, many holding multiple accounts. The incumbents haven't been displaced. Their share of deposits, mortgages, and primary current accounts remains dominant.

What neobanks did was change the reference point against which incumbent performance is evaluated. Before Monzo, the idea that a bank would send you an instant notification after every transaction, with a real-time balance update and the merchant's name formatted as something you'd actually recognize, wasn't something customers expected or demanded — because it wasn't something they had experienced. After Monzo, any bank that did not provide instant notifications was providing an inferior product. The feature itself is not complex. Monzo did not invent real-time payment notifications. But by making them ubiquitous, normalized, and expected — by building a product where they were present from day one, rather than a feature to be added to legacy infrastructure — Monzo permanently raised the floor of what a competitive banking product requires.

The same logic applies to foreign exchange fees, card management, spending categorization, and customer service responsiveness. Revolut built a product that allowed customers to spend abroad without foreign exchange fees and freeze and unfreeze their card from an app. Neither of these features was technologically impossible for incumbent banks. Both had been considered insufficient priorities for the investment they required. Revolut made them table stakes.

This is the neobank's most durable competitive contribution: not the customers they acquired, but the expectations they permanently reset. The incumbents that understood this have responded by building those features into their own products. The incumbents who missed it have found themselves chronically defending a product that their customers are increasingly dissatisfied with.

Nubank in Brazil represents the neobank model's most commercially significant success. The story is worth telling in full because it reveals not just the neobank opportunity but the structural economics of the model at scale.

When David Vélez arrived in Brazil in 2013, he encountered a banking market of extraordinary and durable market power. Five banks dominated the system. Credit card interest rates in Brazil exceeded 300% annually — not 30%, but 300%, a number that's not a typo and is difficult to read without stopping. Bank fees were among the highest in the world relative to income. Customer service was, by international standards, poor. The banks were enormously and consistently profitable — because customers had no alternatives and no realistic prospect of any.

Nubank's founding thesis was elemental: a credit card with no fees, a clean app, and customer service that treated people as adults could build a loyal base. The bank would make money on interchange fees, on credit quality improvements driven by better data, and eventually on the ancillary products that a 100-million-customer base could support. Within eighteen months of launch, 400,000 customers were on the waitlist. Within five years, Nubank had become Brazil's fastest-growing financial institution.

The underlying economics were as important as the product design. Nubank used the granular behavioral data its customers generated to build credit models significantly more accurate than those of incumbents relying on traditional bureau data. More accurate credit models mean lower default rates. Lower default rates mean better unit economics. Better unit economics mean either lower rates for customers or higher margins for the business. Nubank chose both — offering meaningfully lower interest rates than the incumbents, while achieving profitability at scale. By mid-2024, Nubank had surpassed 100 million customers across Brazil, Colombia, and Mexico, reported positive net income in four consecutive quarters, and had demonstrated that the neobank model — patient, data-driven, and customer-centric — could work at a scale that earlier critics had argued was impossible.

THE PARTNERSHIP MODEL: HOW BANKS AND FINTECHS STOPPED FIGHTING AND STARTED BUILDING

By 2020, the most sophisticated banks had reached a conclusion that hadn't been obvious in 2015: the choice between building digital capability organically and competing with fintech challengers was a false binary. The smarter option was partnership — structured relationships in which banks and fintechs exchanged access to the complementary assets each possessed.

Fintechs, in most cases, have superior product design, better data science capability, and faster development cycles. They lack balance sheets, regulatory relationships, distribution at scale, and the deep institutional trust that comes from being a licensed and supervised financial institution. Banks have the balance sheet, the relationships, the distribution, and the trust. They lack product agility, a data science culture, and development velocity.

Partnership structures that allow each party to contribute what it has and access what it lacks have proved significantly more durable than either pure competition or wholesale acquisition.

JPMorgan Chase's engagement with OnDeck Capital — initially a partnership in which OnDeck's credit models were used to assess small business loan applications that Chase would then fund — eventually gave way to Chase building its own small business lending technology in-house, informed by what it had learned through the partnership. This is the partnership model working as its most sophisticated proponents intend: as a mechanism for capability transfer, not just for product distribution.

What these pages establish, taken together, is the nature of the competitive landscape within which financial services are being contested. It isn't a simple narrative of fintechs disrupting banks. It's a complex, multi-directional contest involving platform companies occupying the infrastructure layer, digital challengers resetting expectations, transforming incumbents rebuilding themselves under pressure, and regulatory regimes that simultaneously enable and constrain every actor in the system. Against this backdrop — and with AI about to change the dynamics of all of it — the question of how cash itself is evolving takes on a specific and pointed significance.

Why Bank Tech Migrations Fail — and Succeed

FAILURE	SUCCESS
• **Big Bang Launches** One-time, large-scale overhauls	→ **Incremental Rollouts** Phased changes over time
• **Weak Governance** Limited oversight and accountability	→ **Strong Governance** Clear oversight and accountability
• **Untested Partners** Outsourced to unproven vendors	→ **Proven Expertise** Built in-house or with trusted vendors
• **Sparse Testing** Minimal testing and planning	→ **Continuous Testing** Rigorous testing and planning
• **Compressed Timelines** Rushed rollouts, no rollback plans	→ **Measured Timelines** Gradual scaling with rollback options

Transformations succeed when they are incremental, well-managed, continuously tested, and measured in years — not weeks.

When Piyush Gupta became CEO of DBS in 2009, the bank was, by most assessments, merely adequate. Safe, reliable, and deeply boring. What Gupta understood was that 'adequate' was a deteriorating condition: the direction of digital expectations in Singapore's market meant that a bank that stayed merely adequate would become, progressively, inadequate.

The transformation program launched in 2014 — under the deliberately provocative slogan 'Making Banking Joyful' — was built on four foundations: cloud-native technology setup, data flywheel development, cultural transformation through 'Kill DBS' workshops in which employees designed fintech startups that would put DBS out of business, and an API environment connecting 500+ partner organizations.

By 2022, DBS had been named the world's best digital bank by Euromoney, Global Finance, and The Banker simultaneously — a distinction no other institution had achieved. Its cost-to-income ratio fell from 47 to 40%. Return on equity rose from 10% to the mid-teens. Customer satisfaction transformed from below the Singapore average to consistently above it.

The DBS story is not a story about technology. Technology, as Gupta has said repeatedly, is the easy part. The hard part is the organization — the leadership, the culture, and the willingness to be honest about what needs to change and what must not.

TAKEAWAY *DBS demonstrates that incumbent advantage — balance sheet strength, regulatory relationships, customer trust — isn't an obstacle to digital transformation. In the hands of the right leadership and culture, it's the foundation for it.*

When David Vélez began building Nubank in Brazil in 2013, he encountered a banking market of extraordinary market power. Five banks dominated the system. Credit card interest rates exceeded 300% annually. Customer service was uniformly poor. The banks were enormously profitable — because customers had no alternatives.

Nubank's thesis was elemental: a credit card with no fees, a beautiful app, and customer service that treated people with dignity could build a loyal base. By 2015, 1 million customers were on the waitlist. Within five years Nubank had become Brazil's fastest-growing financial institution, and by 2023 it had surpassed each of the country's incumbent banks in customer numbers.

Nubank used the granular behavioral data its customers generated to build credit models significantly more accurate than incumbents relying on traditional bureau data. Greater accuracy means lower default rates, which means either lower rates for borrowers, higher returns for the lender, or both. Nubank achieved both — offering meaningfully lower rates while achieving profitability at scale.

By mid-2024, Nubank had surpassed 100 million customers across Brazil, Colombia, and Mexico, and was delivering positive net income — vindicating the thesis that the neobank model could be profitable at scale, not just at growth.

TAKEAWAY *Nubank's growth illustrates a dynamic this book traces throughout: when technology enables materially more accurate risk assessment, the most important consequence isn't the reduction of bank profits. It's the extension of credit, at fair prices, to people who were previously excluded from the credit system entirely.*

CHAPTER FOUR

Redefining Cashless

Beyond Notes and Coins

*Cash is more than a payment method. It's a technology
of sovereignty — the one financial instrument that requires
no intermediary, no infrastructure, and no permission.*

To understand the impact of a cashless society on people's everyday lives I have to take you back to my experience during Zimbabwe's 2008 hyperinflation period. I would wake up in the morning, get a bus to the university and by the time it was evening the cost to get back home would have more than tripled. Carrying swads of cash was normal. The central bank kept printing new note denominations until a ten trillion Zimbabwean dollar note was not enough to buy a box of matches. Even today you can find these notes being sold all over the world as the most recent example of financial collapse.

What does this have to do with a cashless society? Sometimes, grasping the true scope of something requires experiencing its complete opposite. Six years later, while living in Shanghai, I found that cash was nearly unnecessary. The habits I'd developed in Zimbabwe—always carrying cash, worrying about getting change at the store, or wondering if the ATM would have enough—felt surreal. In China, everything was cashless. My phone was all I needed. In fact, a single app did it all.

WeChat was the first real 'Super App'. It wasn't just a messaging platform with a digital wallet; it was an ecosystem of services designed to bring digital convenience to almost every major aspect of your life in China. By introducing their robust Mini program API system, the app transcended its normal functionality to become perhaps the only app you would need on your phone. This demonstrated the true power of a cashless society; however, this is not the whole story.

Between 2023 and 2024, a series of brief but revealing outages disrupted critical nodes in Sweden's digital payment infrastructure, affecting systems including the Riksbank's RIX settlement platform, Bankgirot, Swish, and BankID. None lasted long enough to cause widespread economic disruption. But each exposed the same structural fragility: in a country where fewer than 10% of consumers used cash for their most recent transaction and fewer than half of all retailers accepted it, any interruption to digital payment systems left large numbers of merchants and customers with no means of completing a transaction. Sweden had spent a decade routinely dismantling the analogue infrastructure that would have served as a fallback.

Cash.

Sweden is one of the most cashless countries in the developed world. Only 8-10% of Swedes used cash for their most recent payment at the time of the outage, according to the Riksbank's annual payment survey. The ATM network had been shrinking for years. Banks had been closing their cash-handling services. The social and physical infrastructure required to use cash in daily life had atrophied to the point where, for most Swedes in most circumstances, it was simply easier and more natural to pay with a card or a phone.

These outages revealed the structural fragility that Sweden's cashless convenience had created. Sweden had effectively made a societal choice — not through explicit policy but through the accumulated individual choices of consumers, merchants, and banks — to migrate its payment infrastructure entirely to a digital system. The resilience properties of that system depended entirely on its continued functioning. When it failed, even briefly, the economy had no fallback.

In the aftermath of the outage, the Swedish Riksbank published an assessment that attracted more international attention than most central bank reports. Its conclusion was stark: Sweden had allowed its cash infrastructure to deteriorate to the point where the country couldn't function adequately if digital payment systems failed. The Riksbank recommended — and Sweden's parliament subsequently enacted — measures to mandate that banks maintain cash handling services, that certain categories of retailers continue to accept cash, and that the state invest in maintaining the ATM network in rural and underserved areas.

This isn't a Swedish problem. It's the Swedish version of a universal problem: the transition from cash to digital payments creates genuine benefits — lower transaction costs, better fraud protection, greater financial inclusion in markets with high mobile phone penetration — but it does so by replacing a resilient, distributed, infrastructure-independent payment system with one that depends entirely on the continued operation of digital infrastructure that can fail, be attacked, or be made unavailable by deliberate action.

Understanding this trade-off — not as a reason to resist the adoption of digital payments, but as a design constraint that should shape how digital payment systems are built and governed — is the most important single insight in any serious discussion of the cashless transition.

WHAT CASH ACTUALLY IS

The popular narrative of the cashless transition frames it as a story of technological progress: a better payment method replacing an inferior one. This narrative is not wrong. Digital payments are, for most transactions in most contexts, faster, cheaper, safer, and more convenient than cash. But it misses something fundamental about what cash is and what it does.

A banknote isn't simply a payment instrument. It's a direct liability of the central bank — a claim on the state itself. It carries four properties that no private digital payment instrument fully replicates: universality (it must be accepted in settlement of any debt governed by the relevant legal system), anonymity (transactions leave no digital record), finality (settlement is immediate and irrevocable without counterparty risk), and infrastructure independence (it requires no electricity, no telecommunications network, and no functioning software to operate).

These properties aren't historical accidents. They are the result of deliberate design choices made over centuries of monetary evolution, and they serve real functions in a well-designed monetary system. Universality ensures that money circulates without friction through an economy — that no seller can demand a particular form of payment that a buyer can't provide. Anonymity protects legitimate privacy in commercial and personal transactions. Finality eliminates counterparty risk in small transactions, allowing commerce to proceed without the credit checks and settlement delays that would otherwise be required. And infrastructure independence makes cash the monetary system's emergency backstop — the instrument that continues to function when everything else fails.

The cashless transition replaces all four of these properties with something different. Digital payment instruments aren't universal — merchants can decline specific digital payment methods, and people without smartphones, bank accounts, or internet access may find themselves excluded. They aren't anonymous — every digital payment creates a record in at least one institution's systems. They don't achieve finality in the same absolute sense — they are subject to reversal, dispute, and the solvency of the intermediaries that process them. And they are entirely dependent on infrastructure — on power grids, telecommunications networks, software systems, and the continued operation of the companies that provide them.

None of this is an argument against digital payments. The benefits of the transition are real, large, and in most circumstances outweigh these costs. But it is an argument that the transition should be managed thoughtfully — that the resilience and inclusion properties of cash should be preserved either through maintaining cash infrastructure alongside digital alternatives, or through the development of public digital money that replicates cash's essential properties in digital form.

THE BRAZIL MIRACLE: PIX AND THE PUBLIC INFRASTRUCTURE MODEL

The most instructive recent example of how to build a digital payment system that delivers the benefits of the cashless transition without creating the fragilities of the Swedish model is Brazil's Pix — and the story of how it came to exist is as interesting as its remarkable performance.

In 2020, Brazil's central bank, the Banco Central do Brasil, made a decision that was unusual for a central bank and potentially controversial: it wouldn't rely on the private sector to build the real-time payment infrastructure that Brazil's economy needed. Instead, it would build that infrastructure itself, operate it as a public utility, and require every financial institution above a certain size to connect to it. The cost to individuals would be zero. The cost to businesses would be set at a level far below what private card networks were charging. And the participation would be mandatory — there would be no opt-out for banks that preferred to maintain the profitable oligopoly of the existing card payment system.

Pix launched in November 2020. Within its first full year, it processed millions of transactions that nearly eclipsed the total credit card transactions in Brazil. Within three years, it had become the most-used payment method in the country, surpassing cash, debit cards, and credit cards. More than 160 million Brazilians — nearly two-thirds of the population — used Pix by 2023. More than 10 million small businesses were receiving payments through Pix that they had previously received only in cash.

The economics of Pix's success are simple, but their implications are radical. When the infrastructure cost of payment processing drops to near zero, the barriers to merchant acceptance drop with it. A street vendor who couldn't justify the monthly fee for a card terminal, and who previously dealt only in cash, can accept Pix from any customer with a smartphone. A domestic worker who previously could be paid only in cash — creating a paper trail or, often, no trail at all — can receive wages digitally, with a record that supports access to formal financial services for the first time.

The Pix model is the public infrastructure answer to the private platform problem. It provides the same network benefits as WeChat Pay or Alipay — universal acceptance, instant settlement, digital record-keeping — without the concentration of private power that those platforms represent. The data generated by Pix transactions belongs to the customers and institutions involved, not to a private platform that can use it to expand into adjacent financial services on proprietary terms. The pricing is set by the central bank as a public utility regulator, not by a private company optimizing for its own margins.

THE CASH DESERT PROBLEM

While developed economies debate the pace of the cashless transition, a different crisis is developing at the other end of the digital adoption spectrum: the emergence of cash deserts in communities that can't participate in the digital payment economy.

A cash desert, as the term is used in policy research, is an area in which ATMs and bank branches are sufficiently rare that residents face hardship in accessing cash — a problem that cuts across several different populations. Rural communities in the United Kingdom, the United States, and across Sub-Saharan Africa face cash deserts because bank branch closures and ATM decommissioning have outpaced the development of digital infrastructure in their areas. Elderly populations in both developed and developing markets face cash deserts because digital payment systems require literacy, smartphone access, and comfort with digital interfaces that not everyone possesses or can develop. Undocumented immigrants and others without formal identity documents face payment exclusion not because of geography but because digital payment systems require the documentation and banking relationships that cash does not.

The global estimate of people in effective cash deserts — communities where accessing cash is difficult — runs to approximately 1.3 billion. This figure sits awkwardly alongside the headline narrative of the cashless transition, because it suggests that the transition is not uniform and that its costs are disproportionately borne by populations who are already economically marginalized. Designing payment systems with these populations in mind — ensuring that the cashless transition doesn't become a mechanism for financial exclusion — is not a peripheral concern. It's the central test of whether the transition is being managed in the public interest.

FINANCIAL SURVEILLANCE AND THE PRIVACY IMPERATIVE

Any serious discussion of the cashless transition must engage honestly with the dimension that policy debates most often avoid: the relationship between digital payments and financial surveillance.

Cash transactions are anonymous. They require no identification, leave no record, and can't be monitored by any third party. This anonymity serves purposes that most people — if asked directly — would say they value: the ability to make financial choices without being observed, the protection of private spending habits from scrutiny, and the availability of a monetary instrument that can't be weaponized against the holder by political authorities.

Digital payment systems are, by their nature, the opposite. Every digital transaction creates a record — in the payment processor's logs, in the financial institution's systems, and in the data archives of any platform that mediates the transaction. This traceability serves legitimate public purposes: detecting money laundering, identifying terrorist financing, supporting tax enforcement, and enabling the fraud detection systems that protect consumers from criminal exploitation. These are genuine benefits.

But the same infrastructure that serves these legitimate purposes is also, by its nature, a surveillance infrastructure. China's use of digital payment data as an input to its social credit system is the extreme example — a system in which the financial transaction record becomes a continuous assessment of political compliance. But the principles that make this concerning aren't unique to China. The infrastructure required for legitimate anti-money-laundering surveillance is the same infrastructure that could be used, if political conditions changed, for purposes far beyond its stated justification. Every expansion of digital payment penetration expands this infrastructure's reach.

Designing payment systems with privacy as a first principle — building in anonymity protections for small-value transactions as a deliberate design choice rather than an afterthought — is both technically possible and politically necessary. The European Central Bank's Digital Euro proposals include explicit privacy protections for low-value transactions. Several CBDC setups under development use cryptographic techniques that enable transaction validation without exposing transaction details. The design space exists. What it requires is the political commitment to occupy it.

Brazil's Pix, launched by the Banco Central do Brasil in November 2020. Just 2 years later, it processed more transactions than credit cards had processed in all of Brazil — at zero cost to individuals and a fraction of the cost to businesses. The design was deliberately public: the central bank built the infrastructure, set fees, required all banks above a threshold to connect, and operated the system as a utility.

By 2023, more than 140 million Brazilians — two-thirds of the population — had registered for Pix. More than 10 million small businesses were receiving payments through Pix that they had previously received only in cash. Street vendors who couldn't justify a card terminal could accept Pix from any customer with a smartphone.

The Pix model is the public infrastructure answer to the private platform problem. It provides the network benefits of WeChat Pay or Alipay — universal acceptance, instant settlement, digital record-keeping — without the concentration of private power those platforms represent. Data mostly belongs to the customers and institutions involved, not to a platform using it to expand into adjacent financial services on proprietary terms.

TAKEAWAY *Pix demonstrates that the most effective payment infrastructure is often public rather than private — designed for universal access and participation rather than competitive advantage.*

Sweden is one of the most cashless countries in the developed world. Only 8-9% of Swedes used cash for their most recent payment by the early 2020s. Cash acceptance has been declining steadily amongst retailers. The ATM network had been shrinking for years.

When a series of outages between 2023 and 2024 disrupted key components of Sweden's digital payment infrastructure — including the Riksbank's RIX system, Bankgirot, Swish, and BankID — the vulnerability became unmistakable. Merchants who had stopped accepting cash had no fallback payment method to offer customers when digital systems went down.

In the aftermath, the Riksbank published an assessment concluding that Sweden had allowed its cash infrastructure to deteriorate to a point where the country could not function if digital systems failed. Parliament subsequently mandated that banks maintain cash-handling services and that certain categories of retailers continue to accept physical currency.

The Swedish reversal isn't an argument against digital payments. It's an argument for resilience — for ensuring that the cashless transition preserves the emergency backstop function that cash has always provided, either by maintaining cash infrastructure or by using CBDCs that replicate its properties.

TAKEAWAY *Sweden's experience is a warning that the cashless transition, pursued without attention to resilience, creates fragility that is only visible when the digital infrastructure fails — at which point it is too late to rebuild the analogue alternative quickly.*

CHAPTER FIVE

CBDC and the Future of Money

State Power in Digital Form

The issue is not whether central banks should issue digital currency. The puzzle is whether they can afford not to — in a world where private digital currencies are already filling the space that state money is failing to occupy.

130+	13+	$60B	€3,000
Central banks actively researching or developing CBDCs (BIS, 2024)	CBDCs already launched, including Nigeria's eNaira and the Bahamas' Sand Dollar	Destroyed in the TerraUSD algorithmic stablecoin collapse, May 2022	Proposed maximum Digital Euro holding per individual

Personally, I find Central Bank Digital Currencies (CBDCs) a conundrum. Despite all their purported benefits, I can't help but wonder if they are a real solution to a practical problem or an over-engineered response to the rise of digital currencies. In this chapter I will try to resolve that internal conflict.

To many scholars CBDCs are viewed as a means to infuse the benefits of private digital currencies into public money. The promise is that they will enable the public to hold central bank money (reserves) directly. At present, this is only possible for commercial banks and other financial institutions; however, CBDCs are expected to empower businesses and ordinary citizens to use central bank electronic money as a store of value and a means of payment. They are being designed effectively to complement the mandate of central banks, namely maintaining financial and monetary stability. Still, the risks remain. While opportunities exist to fill the cash gap and help in the transmission of monetary policy, the benefits are not clearly defined. Central banks across the globe are at varying stages of CBDC development and the potential to disrupt financial intermediation has become a real threat.

On 25 October 2021, the Central Bank of Nigeria launched a product it had been developing in relative secrecy for two years: the eNaira, Africa's first central bank digital currency. The launch was accompanied by considerable fanfare — live events in Abuja, presidential endorsement, and the kind of technology-forward branding that suggested the central bank had retained consultants who had spent time in Silicon Valley.

Five years on, adoption rates are less than 1%.

This number — one half of one percent of Nigerians using the eNaira after a year of availability — is not primarily a Nigerian story. It's a universal lesson about what central bank digital currencies are, what problems they are designed to solve, and the significant gap between the policy objectives that motivate their development and the user experience that drives adoption.

The eNaira was designed, as most CBDCs are, to solve several problems simultaneously: financial inclusion for the 26% of Nigerian adults without bank accounts, reduction in the cost of government cash distribution, competition for the private digital payment platforms that had been growing rapidly in the Nigerian market, and the establishment of a sovereign digital payment infrastructure that the state controlled directly. These are legitimate objectives. But the design that emerged from them created a product that required a smartphone to use, required bank account verification for higher transaction limits, and offered users essentially nothing that M-Pesa, Paystack, or Opay did not already provide — except the prestige of a government backing that most users did not particularly want or need.

The Nigerian experience is not unique. The Bahamas' Sand Dollar, launched in 2020 as the world's first CBDC, reached fewer than 3% of Bahamian adults in its first three years. Jamaica's JAM-DEX, launched in 2022, has struggled with merchant acceptance. The pattern across these early CBDC implementations is consistent: a technically functional product that solves problems its designers cared about, distributed in an environment that has already developed private digital payment alternatives that users prefer.

Understanding why CBDCs have struggled with adoption in their early implementations, and what design choices would change the outcome, is essential to understand the most fundamental monetary question of our time: what is the future relationship between state-issued money and private digital money?

WHAT IS A CBDC AND WHY DOES IT MATTER?

A central bank digital currency is, at its most basic, a digital form of a banknote. Like a banknote, it's a direct liability of the central bank — a claim on the state itself, not on a commercial bank. Unlike a banknote, it exists only in digital form and can be transacted through digital infrastructure without physical exchange.

This seemingly simple description conceals a set of choices about blueprint, access, interest rates, privacy, and the relationship with commercial banking that determine whether a CBDC is radical infrastructure or expensive irrelevance. The choices are interconnected in ways that make CBDC design one of the most seriously difficult policy problems in contemporary monetary economics.

The foundational question is: who holds the CBDC? If the central bank deals directly with the public — maintaining accounts for individuals and businesses directly on its own infrastructure — then the CBDC is a genuine alternative to commercial bank deposits, with all the structural consequences that implies. If the central bank instead issues CBDC to commercial banks, who then distribute it to their customers, the CBDC looks more like a digital form of the reserves that commercial banks already hold at the central bank — a wholesale rather than retail instrument.

Most serious CBDC proposals have converged on a two-tier model: the central bank issues CBDC and maintains the infrastructure, but commercial banks and regulated payment providers distribute it to consumers and businesses, handling the customer relationships, compliance, and product design. This preserves the existing role of commercial banks in financial intermediation while introducing a new form of money that is backed directly by the state.

The two-tier model solves the disintermediation problem — the risk that direct CBDC access would drain deposits from commercial banks, destabilizing the credit system that those deposits support — but it introduces a different problem: it relies on commercial banks to distribute a product that competes with their own deposit products. Banks have limited commercial incentive to enthusiastically distribute a superior alternative to the accounts on which they earn net interest income. The regulatory question of how to ensure genuine distribution rather than grudging compliance is one that CBDC designers in every major jurisdiction are still wrestling with.

THE DISINTERMEDIATION RISK IN DETAIL

The disintermediation concern is real and deserves more attention than most CBDC enthusiasm gives it. Are CBDCs a solution chasing a problem?

Commercial bank deposits aren't simply a place for households to keep their money. They are the primary raw material of the credit system. Banks fund their loan books with deposits: a €100 deposit at a German bank, subject to reserve requirements and capital ratios, can support several hundred euros of lending in the broader economy. This credit multiplier is how the financial system creates the money supply that the economy operates on.

If a significant proportion of deposit balances migrated from commercial banks to CBDC wallets — because CBDC offered better security, better privacy, or a higher interest rate — the deposit base available to fund commercial lending would shrink. Banks would need to find alternative funding, at a higher cost, for the loan books they were maintaining. The cost of credit would rise. The availability of credit, particularly for smaller borrowers, would contract. Economic growth would slow.

The IMF's modelling of this effect, published in a 2024 working paper, suggests that deposit migration of just 10 to 20% of household deposits to CBDC could reduce bank lending by 3-5% in the short run — a material macroeconomic contraction, particularly if it happened rapidly. The modelling also suggests that the effect is larger in economies with weaker bank capital positions and narrower net interest margins — exactly the environments, in many developing economies, where CBDCs are being considered as tools for financial inclusion.

Central banks have responded to this concern through design constraints: balance caps on CBDC holdings (typically in the range of €3,000 to €4,000 per individual in European proposals), non-remuneration (CBDC pays no interest, eliminating the incentive to hold it in preference to bank deposits), and the two-tier distribution model described above. Each of these constraints reduces the risk of disintermediation. Each also reduces the potential benefits — a non-remunerated, balance-capped CBDC that can only be accessed through a commercial bank app is significantly less radical for financial inclusion than a freely available, interest-bearing digital currency distributed directly by the central bank.

This trade-off — between the disruptive potential of CBDC and the system-wide risk created by its most structural versions — is the central design challenge that no existing CBDC proposal has fully resolved.

THE DIGITAL EURO: THE MOST WATCHED EXPERIMENT

The European Central Bank's Digital Euro project is, by virtue of its scale and institutional weight, the CBDC development most closely watched by the global financial community. A Digital Euro would be a digital form of the currency used by 450 million people across 20 countries — the second most widely held reserve currency in the world. Its design choices would be adopted or adapted by smaller central banks across the eurozone's sphere of monetary influence. And its regulatory regime, developed alongside the EU's AI Act and MiCA regulation, would become part of the world's most far-reaching digital finance governance structure.

The ECB's proposal, as it has evolved through multiple consultation rounds since 2021, envisions a Digital Euro that functions as a digital complement to cash — universally accepted, freely available to any resident of the eurozone, interoperable with existing payment systems, and designed with privacy protections for small-value transactions. It wouldn't pay interest. It would have a holding cap of €3,000 per individual. It would be distributed through commercial banks and payment providers. And it would be accepted by merchants throughout the eurozone at no cost beyond the infrastructure investment of integration.

This design reflects a careful balance between the Digital Euro's stated objectives — providing a European alternative to private digital payment platforms dominated by US technology companies, maintaining cash-equivalent payment options as physical cash use declines, and supporting financial inclusion in communities underserved by private payment providers — and the structural concerns about bank disintermediation that the ECB has made central to its communications.

The privacy provisions deserve particular focus. The ECB's Digital Euro framework includes, for the first time in any major central bank proposal, explicit privacy protections analogous to those of cash for low-value transactions: transactions below a defined threshold would be processed without central bank visibility, using cryptographic techniques that allow the payment to be validated without its content being recorded. Here is a significant design innovation. It amounts to an acknowledgment, at the highest level of European monetary policy, that financial privacy is a legitimate public interest — not simply a preference of those with something to hide — and that the CBDC blueprint should be designed to preserve it.

STABLECOINS: THE PRIVATE ALTERNATIVE

The CBDC debate can't be separated from the stablecoin reality, because stablecoins have been attempting to solve the same problems that CBDCs are designed to address — and in several markets, they have been doing so, chaotically but at scale, for years before any central bank has managed to launch.

A stablecoin is a privately issued digital asset designed to maintain a stable value relative to a reference asset — most commonly the U.S. dollar. At the time of writing Tether (USDT) and USD Coin (USDC) are the two largest, with a combined market capitalization exceeding $150 billion. They are used primarily as a vehicle for cryptocurrency trading — providing a stable unit of account within the volatile crypto environment — but more and more as a means of cross-border payment, a hedge against domestic currency depreciation in high-inflation economies, and in some contexts as a substitute for bank accounts.

In Argentina, where the peso has depreciated by more than 300% against the dollar in the past five years, USDC and USDT have become a widely used savings instrument for middle-class Argentines seeking refuge from inflation. In Venezuela, dollar stablecoins are used to pay for goods and services in communities where the bolivar has lost practical utility. In sub-Saharan Africa, dollar stablecoins sent by migrants represent a faster and cheaper remittance option than traditional wire transfers in several corridors. These are genuine use cases, meeting genuine needs, in markets that the formal financial system hasn't adequately served.

But the stablecoin record also includes the most spectacular failure in the history of digital finance: the collapse of TerraUSD in May 2022. TerraUSD was an algorithmic stablecoin — one that maintained its dollar peg not through dollar reserves but through an algorithmic relationship with a companion token called Luna. The mechanism was mathematically elegant and practically unstable: when confidence in TerraUSD fell, and users began redeeming it for Luna, the supply of Luna increased, its price fell, which further reduced confidence in TerraUSD. This increased redemptions, which further increased Luna supply, in a self-reinforcing collapse that destroyed approximately $60 billion in value in 72 hours.

The TerraUSD collapse established, definitively, the thesis that critics had advanced from the beginning: algorithmic stablecoins are structurally fragile, because their stability depends on market confidence in a mechanism that can't maintain that confidence when tested. The IMF's analysis of the collapse concluded that algorithmically stabilized stablecoins shouldn't be permitted to function as monetary instruments without the full reserve backing that fiat-backed stablecoins provide. This conclusion is now embedded in the regulatory regimes being developed in the EU, the US, and the UK.

The future of money is digital. What matters is which digital money — and whose.

CBDC Design Tradeoffs

DESIGN CHOICE	OPTION A	OPTION B	CORE TRADEOFF
Architecture	**Direct** Central bank issues and manages all accounts and transactions.	**Two-tier** Central bank issues the CBDC; intermediaries (banks/PSPs) distribute and manage it.	Control and simplicity vs. scalability and resilience.
Privacy	**Full anonymity** Cash-like use; no transaction records linked to individuals.	**Traceable** Transactions are recorded for oversight, compliance, and risk management.	Personal privacy vs. transparency and regulatory oversight.
Access	**Universal** Anyone can hold and use a CBDC account.	**Tiered** Basic access for all; higher limits require identification (KYC/ID).	Financial inclusion vs. fraud prevention and risk control.
Offline Capability	**Works offline** Payments work without internet or electricity.	**Online only** Requires internet and compatible devices.	Resilience and access vs. simpler controls and lower cost.
Technology	**Decentralized** Distributed ledger: high resilience and transparency.	**Centralized** Traditional database; easier to manage and update.	Resilience and openness vs. efficiency and control.
Merchant Acceptance	**Mandatory** Merchants must accept CBDC by law.	**Voluntary** Merchants choose to accept, often with incentives.	Faster adoption and reach vs. market freedom and lower burden

No single CBDC design is optimal. The right choice depends on a country's institutions, infrastructure, and the level of trust it seeks to build and maintain.

Research Framework: The CBDC Design Trade-Off Matrix

Every CBDC design involves a fundamental trade-off between four goals: financial inclusion, monetary transmission efficiency, financial stability, and privacy. These four goals can't be simultaneously maximized. An account-based CBDC with full KYC documentation increases compliance at the cost of inclusion for populations who can't provide documentation. A token-based CBDC with transaction anonymity maximizes privacy at the cost of anti-money-laundering effectiveness.

The design space is not binary. Most serious CBDC proposals operate on a tiered model: low-value transactions with lighter requirements and greater anonymity; higher-value transactions with full compliance. The challenge is calibrating the tiers correctly — setting thresholds that provide genuine financial inclusion without creating regulatory arbitrage opportunities that undermine the compliance goals the system must also serve.

The BIS's 2021 survey of CBDC design options and the subsequent research by the IMF's Monetary and Capital Markets Department document the range of technical approaches available and their trade-off profiles. The conclusion is consistent: there isn't a technically optimal CBDC design. There are only design choices that reflect political and policy priorities — and those priorities should be explicit rather than embedded in technical specifications.

Sources: Bank for International Settlements, Central bank digital currencies: system design and interoperability (2021); IMF Working Paper 22/58, How to Issue a Central Bank Digital Currency (2022).

CHAPTER SIX

Fintech and the Global Dividend

Development at Scale

The lesson from mobile money isn't that technology solves financial exclusion. It is that technology, combined with the right regulatory design, the right product architecture, and the right commercial commitment, can solve what has resisted solution for a century.

1.4B	$900B+	15M+
adults globally still without a formal financial account (Global Findex 2025)	global remittance flows in 2025 — outpacing official development assistance	smallholder farmers covered by Pula agricultural insurance across 18 African countries

In his 2014 study "Digitizing the Kaleidoscope of Informal Financial Practices," Ignacio Mas observed that idle balances—where funds aren't invested or saved—were common in digital systems like M-pesa. He identified key features of informal finance that could explain why fintech reaches the unbanked and underbanked but doesn't always drive deeper financial engagement or improve other development metrics. Fintech's promise of financial inclusion would be incomplete if it didn't interrogate why.

Many of the people who had opened M-Pesa accounts, Mas discovered, had done something that seemed rational in retrospect but that the financial inclusion industry hadn't adequately anticipated. They had opened the account. They had received the training. They had made a few transfers. And then they had stopped.

Mas found that many mobile money wallets contained very small or inactive balances— financial capacity that exists in principle but not in practice. The gap between what a person is technically able to do with a financial product and what they actually do with it turned out to be large in many markets. Sometimes the gap was about product design: the product required minimum balances that the user could not maintain, or transaction fees that made small transactions economically irrational. Sometimes it was about trust: the user had had a negative experience — a fee they didn't understand, a transfer that didn't arrive, a customer service interaction that went badly — and had concluded that the formal financial system was not for people like them. And sometimes it was about literacy: the user knew, intellectually, that the account was there, but lacked the confidence to use it effectively, handle its features, or understand what the numbers on the screen meant. Mas suggested behavioral solutions based on digital systems that don't merely digitize informal financial services but understand, adapt and optimize their idiosyncrasies. These suggestions included labelling money for specific goals like school fees in a goal-based savings set up and mimicking social accountability norms that already existed.

This is the transition on which this book's central argument turns. From inclusion to engagement. From access to use. From the first fintech era's most celebrated metric — account ownership — to the metric that actually predicts development outcomes:

whether people are using their financial products in ways that improve their lives.

The distinction is not semantic. It is, arguably, one of the most important steps in understanding the impact of fintech on financial inclusion. And it has direct, significant implications for how the AI-powered next phase of fintech should be designed.

THE ACCESS-ENGAGEMENT GAP: THE DATA

The World Bank's Global Findex survey, conducted every three to four years, is the most far-reaching periodic measurement of financial access and use worldwide. The 2025 edition — the most recent at the time of writing — tells a story of remarkable progress and persistent failure simultaneously.

On the progress side: account ownership has risen from 51% of global adults in 2011 to 79% in 2025 — more than a billion people have entered the formal financial system in a decade. In Sub-Saharan Africa, mobile money account ownership has risen from 12% to 40%. In South Asia, account ownership has risen from 35% to 78%, driven largely by India's Jan Dhan program and UPI environment.

On the failure side: the Findex also reports that approximately 50% of all LMICs account holders —rarely or never use them for active financial management. They hold accounts but don't save in them. They have credit available, but don't use it to invest. They have insurance products they can't understand and can't claim. In Sub-Saharan Africa, the formal savings rate is less than 40% despite account ownership being approximately 75%. The reasons why people don't save or actively use their accounts are nuanced, from a lack of money to a preference for informal methods. When examining the primary factors contributing to this, it is essential to adopt a comprehensive, multifaceted approach.

The author's own research on China, which we have already alluded to in chapter 2 — examining the causal relationship between fintech penetration and financial development across 290 prefecture-level cities — exposes what happens when access does convert to active engagement, and why. The study finds that fintech development has a significant, positive, and causal effect on three dimensions of financial development simultaneously: it expands access to credit, deepens the formal deposit base, and strengthens household savings rates. Cities that experienced greater fintech penetration did not simply open more accounts. They generated deeper financial participation — more lending, more formal saving, and greater household resilience against economic shocks. Crucially, the study also finds that regulatory quality acts as a complement rather than a constraint: cities in provinces with stronger financial governance captured a larger fintech-development dividend than comparable cities with weaker oversight, because better-governed products were more trusted, more actively used, and better designed. The implication for the access-engagement gap is direct. Fintech can close the gap — the China evidence demonstrates this at scale — but only where institutional foundations support active, confident participation rather than passive account ownership.

WHY PRODUCTS FAIL THEIR USERS

The gap between access and engagement is not primarily caused by user failure. It is instead mostly caused by product failure: the systematic design of financial products for populations other than those being served.

Consider the basic savings account. In its standard commercial bank form, the product was designed for a customer with regular employment income, arriving monthly, sufficient to maintain a minimum balance between pay checks. The product requires a minimum balance — to cover the bank's operational cost of maintaining the account — and penalizes the holder when the balance falls below that threshold. The repayment assumption is that the account holder has predictable, regular income.

Now consider the customer profile that dominates newly financially included populations in developing markets: a daily laborer whose income arrives in irregular, unpredictable amounts; a smallholder farmer whose income arrives in two seasonal pulses and is essentially zero for the remaining ten months; a domestic worker paid in cash by multiple employers on different schedules; a street vendor whose daily revenue depends on weather, foot traffic, and dozens of other factors beyond her control. For this customer — and there are hundreds of millions of her, globally — the standard savings product is more than suboptimal. It actively harms her by charging fees for failing to maintain a minimum balance that her income pattern makes structurally impossible to maintain.

The appropriate product design for this customer is radically different. No minimum balance. Interest paid on any balance, regardless of size. Transaction fees — if any — tuned to the size of the transaction rather than to the assumption of sufficient scale. Repayment structures aligned with actual income timing, not with the standard bank's assumption of monthly salary receipt.

These design requirements are known. They aren't difficult to implement on digital infrastructure. They have been demonstrated to work at scale — M-Pesa's savings and credit products were designed exactly on these principles. The reason they're not universal is commercial, not technical: a savings product with no minimum balance and no maintenance fee is a less profitable product than one with minimums and fees, and the commercial incentive to design products that serve low-income users as effectively as they serve high-income users has not, historically, been strong enough to overcome that profitability difference.

AI changes this calculus in a specific and important way. An AI system that can accurately assess which users are likely to become highly active — and therefore highly profitable over a multi-year horizon — even at low initial transaction volumes, can make a genuine commercial case for designing products that serve them well from the beginning. The short-term profitability sacrifice of the no-fee, no-minimum product is offset by the long-term value of the customer relationship, if AI can identify which apparent low-value customers are actually high-potential customers in early stages of financial engagement.

THE GENDER ENGAGEMENT GAP

The gender dimension of the access-engagement gap deserves extended treatment because it is simultaneously the most well-documented and the most under-addressed feature of the financial inclusion landscape.

Globally, women are 4% less likely than men to hold a formal financial account. Among the populations that have entered formal finance through mobile money, women are substantially less likely to progress from basic transfer services to savings, credit, and insurance products. The GSMA Connected Women program documents that women in developing markets are 14% less likely than men to use mobile internet — limiting their access to the digital financial products that are increasingly the frontier of inclusion — and 28% less likely to use mobile money, even in markets where the product is nominally available to them.

The barriers are layered and interact with each other. The first layer is economic: women's lower average incomes reduce their ability to purchase smartphones, maintain connectivity, and sustain the balances that make financial products useful. The second layer is social: in many cultural contexts, male relatives control access to digital devices, financial accounts, and the information required to navigate formal financial systems. The third layer is product design: financial products designed with male primary users in mind — their income patterns, their risk tolerances, their financial literacy levels, their social contexts — may be routinely less suited to women's financial lives and needs.

The Suri and Jack finding on M-Pesa — an approximately 18.5% increase in consumption among female-headed households compared to the roughly 6% average — is often cited as evidence of mobile money's power for gender equity. It is. But it's also worth noting what this finding implies: that the baseline financial exclusion of women is severe enough that the gain from inclusion is proportionally much larger than the average. The 18.5% figure is an indicator not just of inclusion's power but of exclusion's depth.

Research by BFA Global and Women's World Banking has identified the specific product design principles that drive engagement among women in low-income markets. Products must accommodate irregular income patterns rather than assuming monthly salary cycles. Onboarding must work through female social networks and trusted community intermediaries rather than exclusively through formal institutional channels. Customer service must be available in formats that accommodate varying levels of digital literacy. And privacy protections must be robust — the ability to save and transact without a male household member's knowledge or intervention is, for many women, not a preference but a safety requirement.

AGRICULTURAL FINANCE: THE MOST NEGLECTED MARKET

Agriculture employs approximately 30% of the global workforce. It provides the livelihoods of more than half a billion smallholder farming families. It's the economic foundation of most of the world's low-income countries. And it is, by any measure, the most chronically underfinanced segment of the global economy.

The structural reasons are well understood: small transaction sizes that don't justify the fixed costs of traditional credit assessment; dispersed geography that makes physical branch access impractical; seasonal income patterns incompatible with standard monthly repayment schedules; income volatility driven by weather, pests, and commodity prices that makes risk assessment seriously difficult. Traditional banks — designed for urban customers with regular incomes and physical collateral — have no structural advantage in this market and significant disadvantages.

The digital revolution in agricultural finance is beginning to address these barriers in ways that were technically impossible a decade ago.

Satellite imagery, available at commercially viable resolution and update frequency through services like Planet Labs and Maxar, provides crop monitoring data that allows lenders to assess agricultural loan collateral without physical field visits. A lender can now observe, from orbital imagery, whether a borrower's stated maize crop is growing as expected, whether it has been affected by drought or flooding, and whether the harvest is likely to justify the loan that financed the planting. This information — which previously required a physical inspection that was economically impossible for small loans — can now be obtained automatically, continuously, and at a cost that makes it viable even for micro-credit.

Pula, the agricultural insurance startup founded in Kenya in 2015, has built its business on exactly this model. By combining satellite weather data with mobile phone distribution through agricultural input retailers, Pula offers index-based agricultural insurance to smallholder farmers — insurance that pays automatically when satellite data confirms that rainfall in the farmer's area has fallen below an insured threshold, without any requirement for individual loss assessment. The product has been delivered to more than 15 million farmers across 22 African countries. For the vast majority of those farmers, it was their first experience of any formal insurance product.

The evidence is accumulating. A randomized controlled trial of Pula-style index insurance in Ghana found that insured farmers planted on more land, used more inputs, and achieved up to 30% higher yields than uninsured farmers — because the insurance removed the catastrophic downside risk that had been causing them to self-insure through underinvestment. The financial product did not change the farming. It changed the incentive structure within which the farming decisions were made.

THE REMITTANCE REVOLUTION AND ITS LIMITS

Remittances — money sent by migrants to families in their countries of origin — represent one of the largest financial flows in the global economy. In 2025, the World Bank estimated global remittance flows at more than $690 billion for LMICs — more than four times the volume of official development assistance and, in dozens of countries, exceeding both foreign direct investment and export revenues as a source of foreign exchange.

For decades, the economics of international remittance were scandalous by any reasonable standard. As recently as 2015, sending $200 from the United States to sub-Saharan Africa cost an average of 9 to 12% of the transferred amount. That cost — much of it captured by traditional banking intermediaries and cash transfer agencies that dominated the market — represented a de facto tax on the earnings of the world's poorest people's. In some corridors, the fees were higher still.

The digital remittance revolution has driven costs down dramatically in competitive corridors. Wise (formerly TransferWise), founded in 2011 on the principle of genuine transparency about exchange rates and fees, charges 0.2 to 2% for transfers in its most liquid corridors. Remitly, specializing in developing regions, has driven costs in the US-Philippines corridor from more than 8% to under 3%. In Africa, Nala — founded in Tanzania in 2017 — has built a remittance platform specifically for the African diaspora that charges under 2% in several corridors and delivers instantly to mobile money wallets.

The developmental impact of cost reductions in remittances is simple to calculate and substantial. If the cost of sending $200 from the United States to Nigeria falls from 10% to 2%, the recipient receives $196 instead of $180. At the scale of global remittance flows, a reduction from 10% to 2% on average cost represents over $50 billion per year flowing to recipient families rather than to intermediaries. That $50 billion, received by households with median incomes of $2,000, is not a marginal improvement. It's a material addition to household economic capacity.

What this chapter establishes, examined honestly, is that fintech's development impact is real, substantial, unevenly distributed, and contingent. The gains are largest for the most excluded — women, smallholder farmers, informal economy workers — but only where the products are well-designed, the regulatory environment is protective, and the financial literacy exists to make active use possible. Getting that combination right, across the diversity of contexts in which financial exclusion persists, is harder than the headline statistics on fintech adoption suggest. The AI wave now arriving in financial services is the most powerful tool yet built for getting it right. The chapters that follow examine how.

Hamara: When the Value Chain Becomes the Financial Product

In rural Zimbabwe and Zambia, smallholder poultry farmers face a familiar trap: no collateral, no credit history, no access to formal finance. The Hamara Group, headquartered in Bulawayo, Zimbabwe, took a different approach. Rather than building a bank for farmers, it built a value chain around them. Hamara supplies day-old dual-purpose chicks, feed, vaccines, and training. The farmer provides labor and care. At the end of the cycle, Hamara buys back the output through its own abattoirs and distribution network across thirteen cities. Financing is embedded in the contract: inputs are advanced on credit; and repayment is deducted from the sale. No loan application. No KYC. No collateral. The value chain *is* the financial product.

The company has trained thousands of small-scale farmers and operates hatcheries that produce more than 60,000 chicks per week. It is now layering digital tools onto this physical infrastructure — digitizing contracts, tracking production cycles, and generating transaction histories for farmers who've never had one. From the author's direct engagement with the platform, default rates on contract farming arrangements appear remarkably low — a consequence not of vetting, but of structural alignment between inputs, production, and off-take.

The principle: engagement doesn't require a financial product as the entry point. A farmer who receives chicks, feed, training, and a guaranteed buyer is more financially engaged than someone with a dormant mobile wallet — even if she never opens an app. Hamara's model reframes credit risk as a system design problem rather than a borrower-quality problem. Finance the chain, not the individual.

TAKEAWAY *This is an evolving story. Hamara's digital platform is still maturing, and its full impact is yet to be independently measured. But the underlying model — value chain as financial infrastructure — offers a blueprint that deserves close attention.*

CHAPTER SEVEN

The Engagement Imperative

In Zimbabwe today, another fintech story is quietly unfolding. It is the story of InnBucks, a digital wallet service that has embedded itself within one of the country's highest-traffic retail ecosystems — Simbisa Brands. Through outlets such as Chicken Inn, Pizza Inn, and Creamy Inn, the platform has leveraged everyday consumer touchpoints to distribute financial services at scale. InnBucks entered a market long dominated by EcoCash, Zimbabwe's pioneering mobile money platform. Yet it offered something EcoCash had increasingly struggled to maintain: frictionless access to cash through a dense and convenient retail network. In a country where liquidity and the availability of physical cash remain critical features of financial life, that advantage proved powerful.

Established financial institutions have also entered the space. Old Mutual Zimbabwe launched its own digital wallet, O'Mari, positioning it as part of a broader financial services ecosystem. Yet early engagement with such platforms has been more modest than expected. When examined closely, the functional differences between O'Mari and InnBucks are relatively limited. Payments, transfers, and wallet services operate in broadly similar ways. The divergence in user adoption appears instead to reflect something deeper: the strength of the surrounding ecosystem.

Platforms embedded within daily consumer environments — places where people already eat, shop, and transact — enjoy an engagement advantage that product functionality alone cannot replicate. This chapter explores that engagement imperative and examines how artificial intelligence may eventually help financial platforms close the gap. The more difficult question is what that shift might cost in terms of data concentration, behavioral influence, and financial system risk.

We have previously discussed the access-engagement gap. The factors behind this issue are intricate, with no universal solution available. Governments around the world have invested heavily to encourage both financial access and engagement. Fintech companies have introduced various platforms to promote not just inclusion, but also ongoing participation and usage. While some efforts have met with success, the engagement gap remains significant—especially in low- and middle-income countries (LMICs)—and is a critical area of study for anyone interested in fintech. Development agencies have also contributed by funding projects aimed at helping hundreds of millions of people excluded from formal finance, even creating specialized metrics to track progress. Despite these combined initiatives, active financial participation in LMICs remains low.

So why has this investment not translated to sustainable and impactful financial engagement? Financial resilience — measured simply as whether someone can get hold of emergency funds within 30 days — has barely shifted. Accounts were opened and then...nothing. Nobody logged in. The app sat on the third screen of someone's phone, behind the games and the messaging apps, forgotten. I've sat in enough conferences where inclusion numbers are celebrated on stage while practitioners in the hallways quietly admit that the accounts aren't being used. The disconnect is real, and it's uncomfortable.

The access–engagement gap has become one of the central problems in financial development today — more important than the access problem ever was, because it's harder to see and harder to solve. Why does it exist? Some of it is straightforward. As alluded to before, many of the products that were pushed out under the banner of inclusion were miniaturized versions of products designed for wealthier customers. Smaller limits, simpler screens, fewer features — but still built on the assumption that the user has a stable monthly income, a fixed address, and a reason to log in that looks roughly like a middle-class person's reason to log in. That assumption is wrong for most of the people these products were meant to serve. If your income arrives in irregular bursts — a harvest payment here, a day's wages there, a remittance that comes when your cousin in the city has a good month — a savings product that expects a monthly deposit isn't useful. It's irrelevant. And people don't engage with irrelevant things, no matter how many SMS reminders you send them.

But there's a layer beneath the product design problem that's more fundamental. People engage with financial tools when those tools solve a problem they can feel right now. Not a problem they've been told they have. Not a problem that shows up in a policy document. A problem that they are facing today. Can I pay for my child's bus fare? Can I buy airtime without walking to a kiosk? Can I check whether the money my sister sent has arrived? If the answer to those questions is "yes, open this app," then the app gets opened. If the answer is "no, this app just holds your account balance," then it doesn't.

Ozili saw this coming back in 2018. In a paper in the Bursa Istanbul Review that hasn't received nearly enough attention, he argued that digital finance only works as an inclusion tool when it's built for the actual rhythms of low-income life: small-value, high-frequency transactions. Not scaled-down banking. Something genuinely different. He also made an observation that challenged the prevailing orthodoxy: people willingly pay more for digital services when the convenience is real. They're not just price sensitive. They're relevance sensitive. Give them something that fits, and cost becomes secondary. Give them something that doesn't fit, and even if it's free, it will remain unused.

CGAP's analysis of the 2025 Findex data brought this home for me in a way that's hard to argue with. East Asia and South Asia have arrived at almost identical levels of account ownership. But when things go wrong — a health crisis, a death in the family, a failed harvest — the proportion of people who can actually access emergency funds in East Asia is more than double that of South Asia. I've turned that comparison over in my head many times. Nearly the same accounts. Entirely different lives. What's different is not whether people have financial tools. It's whether those tools are part of their week — part of how they pay, save, borrow, and cope. That's engagement. And that's what this chapter is about.

What Alipay and WeChat Actually Got Right

There's a version of the Alipay and WeChat story that is widely publicized at conferences and in policy papers, and it goes roughly like this: China's government got behind mobile payments, the market was enormous, credit cards never took off, and adoption followed naturally. It's not entirely wrong. But it's the kind of explanation that sounds satisfying without explaining anything. It tells you what happened. It doesn't tell you why.

Alipay didn't begin as a payments company. It began as a solution to a problem that nobody in the financial inclusion world was thinking about; trust in online commerce. When Alibaba's marketplace was young, Chinese buyers had no reason to trust Chinese sellers. You couldn't inspect the goods. You couldn't return them easily. And if the seller took your money and vanished, you were stuck. Alipay was an escrow service. It sat between buyer and seller, held the funds, and released them only once the goods arrived. It solved a commerce problem. The payment was incidental. Nobody thought of it as "finance."

That's the key. And I want to dwell on it because it explains almost everything about why engagement on these platforms is so much deeper than anything the inclusion movement has achieved. Alipay didn't ask anyone to change their relationship with money. It didn't require anyone to become "financially literate." It just made shopping safer. The money moved as a by-product of buying things. And by the time the company pivoted to mobile payments, it had the trust of an enormous user base that had never once thought of Alipay as a financial service.

WeChat Pay came at the same problem from the other direction. Tencent's WeChat was a messaging platform — essentially China's WhatsApp, except that it became far more than that. Payments were bolted onto social behavior. The masterstroke was the Red Packet feature: a digital version of the traditional Lunar New Year custom of gifting cash in red envelopes. You could send a red packet to a group chat and let people scramble for a random share. It was a game. It was a tradition. It was the kind of thing people forwarded to their parents, who forwarded it to their parents. And every red packet that was opened deposited a small amount into a WeChat Pay wallet. Millions of wallets got seeded with initial funds not through a marketing campaign, but through a cultural moment. Nobody signed up for a "digital wallet." They rang in the New Year.

Once the money was in the wallet, behavior shifted. If you've got a balance sitting in an app you already open twenty times a day for messaging, it's a very short step to using it to pay for lunch. Or a taxi. Or utilities. The QR code system made this frictionless for merchants too — a printed code on a piece of cardboard, no hardware needed. Street vendors adopted it. Market stallholders adopted it. It spread through the economy not because anyone told people to go cashless, but because it was easier than fumbling for change.

What connects these two stories is a principle that the financial inclusion world has been slow to absorb: engagement doesn't come from making financial services available. It comes from making them invisible. When finance is the main thing you're offering, users treat it as a chore. They engage reluctantly, infrequently, and only when they have to. But when finance is buried inside something they already want to do — shopping, chatting, gifting, eating — it becomes part of the texture of daily life. Clearly, engagement is a function of convenience. When fintech platforms offer services that simplify the lives of their customers, that's when the magic happens. We already knew this, but why is it so hard for finance to apply this basic principle in LMICs? Is it because maybe offering this convenience is not a profitable or financially sustainable approach relative to the status quo?

This is the part that really matters for policymakers trying to learn from the China case. The Chinese government didn't orchestrate this. For most of the past decade, regulators were doing something different: trying to contain the Alipay–WeChat duopoly. The suspension of Ant Group's IPO in 2020. The forced interoperability between the two ecosystems, tearing down the walled gardens both companies had carefully built. These were not acts of encouragement. They were a government reasserting control over payment infrastructure that two private companies had effectively captured.

The engagement was organic. The regulation was reactive. If you're a policymaker in Nigeria or Indonesia or Brazil trying to figure out how to close your own engagement gap, that distinction should change how you think about the problem. You can't mandate engagement into existence. But you might be able to create the conditions for it — and the most important condition turns out to be letting financial services embed themselves into activities that people already care about, rather than standing alone as products that people are supposed to care about on their own.

How AI Could Fill the Gap

Most countries aren't going to produce their own Alipay. I want to be realistic about that. The conditions that made China's super-app duopoly possible — the e-commerce backbone, the social media reach, the low credit card penetration, and the years of light-touch regulation that gave both platforms room to grow unchecked — were specific to a time and place. You can't copy-paste an ecosystem. So, the question is more practical than that. If deep engagement requires embedding finance into daily life, and if you don't have a super-app to do it, what else is there?

Artificial intelligence is part of the answer. Not AI as a brand exercise or a press release, but AI as the thing that makes financial products feel personally relevant to people whose lives don't look like the lives those products were designed for. Think about a woman running a small trading business in Harare. Her income comes in unpredictable amounts on unpredictable days. She knows she should be saving — for her children's school fees, for the slow season, for the unexpected. But every time money hits her account, there are things that need to be paid right now. By the end of the week, the balance is gone. A traditional savings product would tell her to deposit a fixed amount on the first of every month. That's useless to her. She doesn't have a "first of the month."

Now imagine a system that watches her transaction patterns — not in a surveillance sense, but in the way a good bookkeeper would — and notices that she tends to receive payments on Tuesdays and

Fridays, and that her spending accelerates on Wednesdays and Saturdays. It could auto-sweep a small amount into a ring-fenced pot each time a payment arrives, before she has a chance to allocate it elsewhere. Label it "School Fees." Make it slightly awkward to withdraw from. She didn't have to set it up. She didn't have to remember. The system did the remembering for her, and it did it in a way that mirrors what she'd do herself if she had the time and the mental bandwidth — the envelope under the mattress, the cash left with a neighbor, the contribution to the savings group.

That's personalization that fits. It's not sophisticated in a technical sense. The machine learning involved is relatively straightforward. What's hard is getting financial institutions to think this way — to start with the user's life and work backwards to the product, rather than starting with the product and hoping the user's life conforms to it.

Then there's credit. This is where AI has arguably had the most visible impact already. In economies where most adults don't have formal credit histories — no payslips, no utility bills in their name, certainly no collateral — the traditional lending model simply doesn't work. Machine learning trained on alternative signals can approximate creditworthiness in ways that weren't possible a decade ago: mobile top-up patterns, bill payment consistency, and how someone navigates their phone. It's imperfect, and I'll get to the risks shortly. But it opens a door that was closed before. And credit, once established, is an extraordinarily powerful engagement anchor. A person with an active loan and a growing repayment history has a reason to stay on the platform. Their track record becomes an asset — one that exists only on that platform and can't be transferred elsewhere. That's lock-in, and it cuts both ways, but it's undeniably engagement.

The piece that excites me most, though, is automated resilience. Parametric insurance that triggers when satellite imagery detects a drought. Crop assessments done from a phone camera. Financial guidance delivered in Xhosa or Tonga through a conversational interface that actually understands context, not just vocabulary. These capabilities are still patchy in practice. But where they work, they do something remarkable: they give a dormant account a reason to wake up. The payout arrives in the wallet. The nudge appears after a deposit lands. The insurance claim resolves itself. Each one of these is a moment where the account becomes useful — and a user who experiences the account as useful is a user who comes back.

The promise isn't that AI replaces the human relationships that make finance work in low-income communities. It's that it extends the reach of useful financial services beyond what any human-mediated system can cover. A loan officer in a branch can assess a few dozen applications a week. A well-designed model can assess millions. That's not a qualitative improvement. It's a different category of possibility.

The Tradeoff: Engagement at the Cost of Stability?

I want to be honest about what worries me.

Everything I've just described — the personalization, the alternative credit scoring, the automated insurance — sounds like progress. And in many ways, it is. But each of these mechanisms also introduces a new kind of fragility into financial systems, and I don't think the development community has honestly reckoned with that.

Start with personalization. When one algorithm is nudging millions of people at the same time — auto-sweeping savings, adjusting credit limits, recommending products — and that algorithm has a flaw, the flaw doesn't stay local. It doesn't just affect a single customer at a single branch. It propagates across an entire market at the speed of a code deployment. Traditional financial systems have a kind of messy, inefficient resilience built into them: different loan officers in different branches making different judgement calls, some good, some bad, but in aggregate creating a diversified portfolio of decisions. AI-powered systems replace that diversity with uniformity. Every user gets the same model. And when the model is wrong, everyone gets the same wrong answer at the same time. That's correlated risk, and it's the kind of risk that turns problems into crises.

Alternative credit scoring makes this sharper. A model trained on historical data will reproduce whatever patterns that data contains — including the biases. If informal workers have historically been excluded from credit, the model learns that informal work is a risk factor. It doesn't know that the exclusion was unjust. It just knows the pattern. The result is a system that excludes the very people it was designed to include, but does so behind a confidence score and an automated notification that says "your application was unsuccessful." No explanation. No recourse. Digital finance, left ungoverned, could undermine the stability gains that inclusion was supposed to deliver. What's different now is the scale at which it can happen.

There's something I find particularly troubling about this. A woman in Bulawayo who has been turned down by an algorithm isn't told that her income pattern was flagged as irregular. She gets a rejection notice. If three platforms reject her, she doesn't think "the models are biased." She thinks "formal finance isn't for people like me." And from the data's perspective, she appears to be someone who chose not to engage. Voluntary exclusion. Case closed. Except it wasn't voluntary. She was pushed out by a system that couldn't see her clearly.

Then there's concentration. When AI-driven engagement makes one platform dominant in a market, that platform stops being a company and starts being infrastructure. And infrastructure that fails doesn't just affect shareholders. It affects everyone who depends on it. China's peer-to-peer lending sector taught this lesson brutally. Platforms used technology to extend credit at breathtaking speed — celebrated as "democratizing finance" right up to the moment they collapsed, taking the savings of ordinary families with them. The technology that made the growth possible was the same technology that made the collapse catastrophic. Speed works in both directions.

I also worry about something subtler. The features that make a fintech app "engaging" — one-tap loans, frictionless spending, instant credit top-ups — are the same features that make it easy to overextend. Friction in financial systems isn't always a bug. Sometimes it's a feature. The moment of hesitation before you take out a loan. The inconvenience of walking to a branch. The time it takes to fill out a form. These small frictions give people space to reconsider. When AI removes all of them in the name of engagement, we should ask whether we've also removed the guardrails.

So where does this leave us? With a genuine tradeoff, not a hypothetical one. AI-powered fintech can close the engagement gap. It can turn dormant accounts into active financial lives. But the same capabilities that make that possible also concentrate risk, encode bias, and create single points of failure in already fragile systems. The question for policymakers isn't whether to permit AI in finance — that decision has already been made by the market. The question is whether governance can keep pace.

In most emerging markets, I don't think it can. Not yet. Financial regulators are set up to supervise banks: institutions with balance sheets, periodic reporting, and prudential frameworks built over decades. Fintech platforms that use AI to make real-time decisions across multiple product categories, growing faster than any supervisor can track — these don't fit that framework.

The governance gap between what technology can do and what institutions can oversee may be the most consequential gap in this entire space. More consequential, I'd argue, than the access–engagement gap this chapter started with. Because if we close the engagement gap and blow up the financial system in the process, we haven't made progress. We've just created a more sophisticated form of failure.

Case Study: Why Revolut Achieved Deep Engagement

Revolut is a London-founded fintech that serves a mostly European customer base. It's not solving for the unbanked. Its users are young, urban, digitally fluent — worlds apart from the smallholder farmer or the market trader we've been discussing. But the design principles that explain Revolut's engagement are the same ones that explain Alipay's and WeChat's. And understanding them in a Western context helps strip away the assumption that deep engagement is somehow a uniquely Chinese phenomenon tied to unique Chinese conditions.

Revolut started in 2015 with a single product: a travel card that offered interbank exchange rates without the markups charged by traditional banks. That was the hook. Not "better banking." Cheaper holidays. Nik Storonsky, the founder, was a former trader who was personally annoyed by FX fees. He built a product for people like himself. Travelers downloaded the app, loaded funds, used it abroad, and then — and this is the part that matters — came home and kept using it. The app was already on the phone. The money was already in the wallet. Switching back to Barclays or HSBC meant downloading another app, re-entering card details, and relearning an interface. The path of least resistance was to stay.

That's the entry point. What followed was a relentless expansion of reasons to return. Budgeting tools that categorize spending automatically. Savings pots that round up every purchase to the nearest pound and sweep the difference into a vault. Stock trading. Crypto. Insurance. Business accounts. eSIMs for travel. Each feature gave users another reason to open the app, and — crucially — each interaction generated data that made the next feature more relevant to that specific user.

By Revolut's own numbers, the average user now transacts dozens of times a month. Think about what that means. That's not someone who checks their balance occasionally. That's someone for whom the app is part of the daily rhythm of spending, saving, and managing money. It's a habit, not a service.

What did Revolut get right that traditional banks, with all their resources and customer relationships, have struggled to replicate? Revolut organized around what people do, not what banks sell. A traditional bank is structured in product silos: the savings team, the lending team, the insurance team, and the investments team. Each silo has its own onboarding, its own interface, its own customer journey. The experience of being a customer is the experience of navigating between fiefdoms. Revolut collapsed all of that into a single surface organized around actions: spending, sending, saving, investing. The difference sounds cosmetic, but it's architectural. When the structure follows behavior, adopting the next product costs almost nothing — no new app, no new KYC, no new learning curve. And multi-product adoption is the single best predictor of sustained engagement in digital finance.

Then there's the way Revolut made financial management feel like something other than homework. Savings challenges with progress bars. Real-time spending summaries that arrive as push notifications, not as monthly statements nobody reads. Gamified onboarding that carries new users through compliance without making them feel like they're filling out tax forms. I'm generally skeptical of gamification as a concept — it's overused and often patronizing — but Revolut's implementation is restrained enough to work. It makes small financial tasks feel like small wins, and small wins build habits.

Revolut also built virality into the product in a way that's genuinely clever. Most of its new users come through word of mouth and referrals, not paid advertising. That's not an accident. Fee-free peer-to-peer transfers mean that every time a Revolut user splits a bill with someone who doesn't have the app, the non-user sees how seamless it is. Curiosity and mild social pressure do the rest. It's the same mechanic that powered WeChat's Red Packets, adapted for a Friday night dinner in London. Different culture, identical engine.

And finally — the piece most relevant for the broader argument of this chapter — Revolut closed the feedback loop. A traditional bank tells you what happened to your money at the end of the month. Revolut tells you what's happening to your money right now. Real-time notifications when you spend. Instant balance updates. AI-generated observations: "You spent more on restaurants this week than all of last month." The information arrives while you can still change your behavior. That's the engagement layer that AI enables at its most practical. Not a prediction. Not optimization. Just interpretation, delivered when it matters.

I dwell on Revolut not because it's a model that can be transplanted directly into Dhaka or Nairobi. It can't. The user base, infrastructure, and regulatory environment are different. But the underlying logic is transferable, and it's the same logic that explains every engagement success story in this chapter.

Finance is not the destination. It's the infrastructure for something the user already cares about. Travel, in Revolut's case.

Shopping, in Alipay's. Socializing, in WeChat's. When finance is the headline, you build accounts. When finance is the plumbing, you build habits. And habits — not accounts — are what turn inclusion into development outcomes.

From Counting Accounts to Building Lives

This chapter has covered a lot of ground, but the argument running through it is simple enough to state in a few sentences. The financial inclusion movement's greatest achievement — the near-doubling of account ownership in low- and middle-income countries — is also its most revealing limitation. Access was necessary. But access without engagement is a hollow victory. An account that sits dormant doesn't build resilience. It doesn't smooth consumption when the harvest fails. It doesn't help a family absorb a medical emergency. It's a number in a progress report, and the people whose names are attached to those numbers deserve more than that.

The platforms that cracked the engagement problem — Alipay, WeChat Pay, Revolut — didn't do it through government mandates or awareness campaigns. They did it by embedding financial services into activities that users already cared about. They are designed for behavior, not for product categories. They made the financial product invisible, and engagement followed as a natural consequence. No one had to be convinced. The product was simply there, woven into the fabric of shopping, messaging, travelling, and living.

AI has the potential to bring this same logic to markets that lack a dominant super-app. Personalization at scale. Credit for people without credit histories. Insurance that pays out without paperwork. Guidance in languages that the formal financial system has historically ignored. These capabilities can convert dormant accounts into active financial lives. But the mechanisms that make

AI effective also introduce fragilities: correlated risk, encoded bias, platform concentration, and a governance gap that most emerging-market regulators aren't yet equipped to close.

That tradeoff won't resolve itself. It requires choices — about regulatory design, about data governance, about who these systems are ultimately built to serve. Those choices are still being made, in some cases by people who understand the technology and in many cases by people who don't. Getting them right matters more than any algorithm.

Inclusion was the first act. Engagement is the second. And it's the act that determines whether a decade of progress translates into better lives — or remains, at its worst, a very expensive counting exercise.

CHAPTER EIGHT

Too Big to Fail Fintech

Systemic Risk in New Clothes

The next financial crisis won't look like the last one. It never does. But it will share one essential feature: institutions that grew so inter-connected that their failure became unthinkable — until it happened.

$315B	€24B	20	$32B
Ant Group's valuation at time of suspended IPO — larger than JPMorgan Chase	in market value wiped out within days when Wirecard filed for insolvency	fintech firms classified as 'systemically important' by at least one major regulator	value destroyed in the FTX collapse, November 2022

On 3 November 2020, the most anticipated technology IPO in Chinese history was scheduled to begin trading on the Hong Kong and Shanghai stock exchanges. Ant Group, the financial affiliate of Alibaba, had priced its shares at a level that would have valued the company at $315 billion — making it, at the moment of listing, more valuable than JPMorgan Chase, the largest bank in the United States.

Two days before trading was due to begin, the listing was suspended. Chinese regulators had summoned Ant's founder, Jack Ma, to a meeting. They had reviewed the prospectus and found something that, in their assessment, it transformed what had been presented as a technology company into something else entirely: a systemically important financial institution operating without the capital requirements, the supervisory oversight, or the resolution planning that systemic financial importance demands.

Ant Group was not a bank. It had always insisted, with some legal justification, that it was a technology company that happened to facilitate financial transactions. It did not hold deposits in the traditional sense. It did not carry credit risk on its balance sheet because the loans it arranged were funded by its banking partners. It was not subject to bank capital requirements because it was not classified as a bank. And yet, as Chinese regulators had calculated, its micro-lending platform had extended credit to more than 500 million individuals. Its payment system processed more daily transactions than Visa and Mastercard combined. Its money market fund managed assets exceeding those of most central bank reserves. The failure of Ant Group — whether through a credit crisis, an operational collapse, or a run on its money market products — would have consequences for the Chinese financial system that no regulator could be indifferent to.

The Ant Group suspension was, in retrospect, the moment that the too-big-to-fail problem in fintech became undeniable. The irony was almost too neat to have been designed: the fintech revolution, which had been celebrated in its early years as the solution to the concentrated, interconnected, systemically dangerous banking system that had produced the 2008 financial crisis, had created a new generation of institutions whose systemic importance was, if anything, larger — and whose regulatory regime was, if anything, thinner.

THE ANATOMY OF FINTECH SYSTEMIC RISK

To understand the system-wide risk that fintech has introduced into global finance—it helps to separate it into its distinct forms — because the risks are different in character from those of traditional banking, require different regimes to manage, and in some cases require regulatory tools that don't yet exist.

The first form is infrastructure concentration risk. The financial system has migrated a significant and growing share of its critical operations to a small number of technology infrastructure providers. Amazon Web Services, Microsoft Azure, and Google Cloud collectively host an estimated 70% of global financial services workloads. This concentration creates a category of systemic dependency with no historical precedent in financial regulation: not the failure of a bank, whose impact is bounded by its balance sheet, but the failure of a layer of shared infrastructure on which hundreds of financial institutions concurrently depend.

A major outage at Amazon Web Services doesn't affect only companies that run on AWS. It affects every company whose counterparties, payment processors, data providers, and market infrastructure also run on AWS — which, in an interconnected financial system, means nearly everyone simultaneously. The 2017 AWS us-east-1 outage, which lasted four hours and affected thousands of websites and services, provided a preview at a modest scale of what a severe cloud infrastructure failure could mean for financial markets. Several fintech payment processors went offline. The cascading effect across their merchant clients and those merchants' banking counterparties illustrated, in miniature, the systemic potential of cloud concentration failure.

The EU's Digital Operational Resilience Act — DORA — which became fully enforceable in January 2025, represents the most serious regulatory attempt yet to address this risk. DORA requires financial institutions to identify their critical technology providers, assess their resilience, maintain oversight of those providers' operational standards, and — most significantly — establish that critical third-party providers are themselves subject to supervisory oversight by European financial regulators. This last provision is the conceptual breakthrough: it extends the perimeter of financial regulation beyond financial institutions to the technology companies on which those institutions depend. It establishes, for the first time, that systemic importance is a regulatory concept that applies to infrastructure providers, not just to deposit-taking institutions.

DORA's limitations are practical rather than conceptual. Enforcing its third-party oversight provisions requires financial regulators to develop the technical expertise in assessing cloud infrastructure resilience — a capability most financial supervisors don't currently possess. Its geographic scope is limited to the EU, while the technology providers it seeks to govern are global, raising questions about enforcement against providers with limited EU presence. And the supervisory resources required to engage meaningfully with hyperscale technology companies — organizations whose annual revenues exceed the GDP of many EU member states — represent a challenge to regulatory capacity that hasn't yet been adequately addressed.

The second form of systemic risk is platform dependency at the application layer. As payment processing, credit origination, and financial data infrastructure have migrated to a small number of fintech platforms, the operational dependencies of the broader financial system on those platforms have grown in ways that are inadequately mapped by existing supervisory regimes.

Stripe, the payments infrastructure company, provides the most instructive case. Founded in 2010 and built to process online payments for internet businesses of any size, Stripe grew from a developer tool to a foundational layer of digital commerce infrastructure through the 2010s. By 2022, Stripe was processing payments for several million businesses in over 40 countries, including a significant share of the world's most economically significant internet companies — from Amazon to Salesforce to Shopify. The majority of global e-commerce, by some estimates, touched Stripe's infrastructure at some point in the transaction chain.

A Stripe outage is no longer merely an inconvenience for the companies affected. It is an economic event with material consequences for employment, government tax revenues, and consumer welfare across multiple jurisdictions simultaneously. The concentration of financial infrastructure in a private company with no regulatory capital requirements, no resolution framework, and no public accountability for operational continuity represents precisely the kind of systemic importance that Prudential Financial regulation was designed to prevent — and that the regulatory regimes built before Stripe's existence weren't designed to catch.

The FCA's analysis of the UK fintech sector documents this dependency in concrete terms: more than 60% of UK financial services firms depend on fewer than five critical technology vendors for core operational functions. When one of those vendors fails, the cascade potential across the financial system is significant. And the supervisory regimes designed to manage bank failure have no obvious mechanisms to manage the failure of a technology platform on which multiple banks simultaneously depend.

The third form of systemic risk operates through credit cycle dynamics. As fintech lending has grown from a niche to a mainstream component of credit supply in several markets — accounting for more than 50% of all unsecured loan debt in the United States— the question of how fintech lenders behave through a full credit cycle has moved from an academic concern to a regulatory imperative.

The standard fintech lending model — originate to distribute, in which platforms make lending decisions and then sell the resulting loans to institutional investors rather than holding them on the balance sheet — creates an incentive structure that looks uncomfortably familiar to anyone who studied the US mortgage market in the years before 2008. When loans are originated but not retained, the originating platform bears the reputational risk but not the financial risk of poor credit decisions. The institutional investors who purchase the loans bear the credit risk but have limited visibility into the origination standards that produced it. The resulting information gap is identical, structurally, to the asymmetry that allowed US mortgage originators in 2005 and 2006 to dramatically lower their underwriting standards because the consequences of those decisions would be borne by investors who had bought their mortgages through securitization vehicles.

The COVID-19 pandemic provided a partial stress test. Several prominent US fintech lenders — LendingClub, Prosper, Kabbage — substantially contracted their originations or sold their portfolios when consumer credit stress rose in early 2020. The platforms that had marketed themselves as superior credit models retreated from the market at precisely the moment the credit cycle turned, leaving borrowers without the credit access that had become part of their financial planning. The resilience of the originate-to-distribute model under genuine macroeconomic stress remains unproven at the scale at which it has since grown.

THE RESOLUTION PROBLEM

Traditional banking regulation has, over decades of crisis experience, developed a sophisticated framework for managing the failure of systemically important banks. Resolution regimes allow banks to be restructured in an orderly manner, preserving the critical financial services they provide while imposing losses on shareholders and creditors rather than taxpayers. Bail-in tools convert debt to equity, recapitalizing a failing bank without public funds. Living wills — mandatory resolution planning documents — require banks to maintain internal structures that make orderly failure possible.

These structures are imperfect. No major economy has yet tested its resolution regime against the simultaneous failure of multiple globally systemic banks. But they exist, they have been developed with considerable regulatory intelligence, and they provide a governance structure within which supervisors can act with legal authority when systemic banks get into trouble.

No comparable framework exists for the resolution of a systemically important fintech platform. The failure of Ant Group wouldn't be managed by a resolution regime designed for banks. The failure of Stripe would not be addressed by a banking resolution authority with the statutory powers and tools needed to stabilize a technology company mid-failure. The failure of a major cloud provider hosting financial industry workloads would likely fall almost entirely outside any existing resolution framework — because no such framework was designed with that scenario in mind.

This is more than an oversight of regulatory design. It reflects a genuine conceptual challenge: resolving a software platform is categorically different from resolving a bank. A bank's balance sheet contains identifiable assets and liabilities that can be separated, valued, and transferred to an acquiring institution. A fintech platform's value resides in its code, its data, its contractual relationships, and its network effects — none of which can be plainly transferred in the way that a bank's loan portfolio and deposit book can be.

THE CORRELATED AI RISK

The most novel and least well-understood form of systemic risk in the AI-powered financial system is also the hardest to regulate: the correlated behavior risk that emerges when multiple large financial institutions rely on similar, or in some cases identical, AI systems.

When every major bank's credit model is built on a variation of the same foundational architecture, trained on similar datasets by data scientists who have studied the same research literature, updated on similar schedules by vendors operating similar review processes — the diversity of credit behavior that historically provided financial system resilience begins to decline. If a flaw in the shared architecture causes all these models to simultaneously misestimate a specific kind of credit risk, the error is not diversified away by the fact that multiple institutions independently make it. It is amplified because multiple institutions are making it in the same direction at the same time.

The 2010 Flash Crash — in which algorithmic trading strategies that shared similar pattern recognition approaches collectively triggered a self-reinforcing sell-off that erased approximately a trillion dollars of market value in minutes — is the prototype for what correlated AI failure looks like. The Flash Crash was recovered within hours, because the underlying asset values had not changed, and rational participants could identify the overshoot. A correlated AI failure in credit markets — where the errors accumulate over months or years of lending decisions before the cycle turns — may be harder to fix when eventually detected.

The IMF, in its 2025 Global Financial Stability Report, identifies correlated AI behavior as one of the major primary systemic risk channels introduced by AI in finance, alongside model opacity and decision speed. The policy response — diversifying AI architectures across institutions, requiring documentation of model architecture choices, stress-testing credit models against scenarios designed to reveal their shared blind spots — is technically demanding and not yet implemented at the scale the risk warrants.

CASE STUDY — THE 2022 CRYPTO CONTAGION: WHAT SYSTEMIC RISK LOOKS LIKE IN PRACTICE

The sequence of events in cryptocurrency markets from May to November 2022 — from the TerraUSD collapse through the implosion of Three Arrows Capital, Celsius Network, Voyager Digital, BlockFi, and FTX — provides the clearest available real-world illustration of how interconnection, opacity, and leverage combine to create systemic contagion in a lightly regulated financial system.

The TerraUSD collapse, described in Chapter Five, destroyed approximately $60 billion in value in a matter of days in May 2022. But the collapse did not end there. Three Arrows Capital, a crypto hedge fund that had taken significant leveraged positions in Terra and Luna, was unable to meet margin calls and defaulted on loans from multiple crypto lending platforms. Those platforms — Celsius, Voyager, BlockFi, Genesis — had in turn extended credit to retail customers, used customer deposits to fund their lending, and structured their balance sheets with opacity that prevented the customers who had entrusted them with funds from understanding the risk they were bearing.

When Three Arrows Capital defaulted, the lending platforms that had extended it credit experienced simultaneous losses that exceeded their capital. They suspended withdrawals — preventing customers from accessing funds they had deposited— on the reasonable belief, based on the marketing materials but incorrect in practice, that those funds were safe. Within months, all four had filed for bankruptcy. The retail customers who had deposited funds — many of them attracted by the double-digit yield promises that the platforms had marketed — suffered losses that, in aggregate, exceeded $10 billion.

FTX's collapse in November 2022, driven by the revelation that the exchange's sister trading firm, Alameda Research, had been using customer deposits to fund its own trading positions, destroyed an additional $32 billion and cost its customers — including substantial institutional depositors — an amount that subsequent bankruptcy proceedings have slowly been recovering. The FTX story was not primarily a technology story. It was a fraud story. But the technology — the opacity of blockchain accounting, the speed of digital fund transfers, the global regulatory arbitrage that allowed FTX to operate from the Bahamas while serving customers in more regulated jurisdictions — was what made the fraud possible at the scale it reached before it was detected.

The crypto contagion of 2022 did not produce systemic consequences for the broader financial system because, at that point in time, the interconnections between the crypto landscape and the traditional financial system were limited. Banks had relatively modest direct exposures. Institutional investors had allocated small proportions of their portfolios. The regulatory structures that would have been required to manage a worse outcome — clear custody rules for digital assets, capital requirements for crypto intermediaries, resolution structures for crypto exchanges — were absent, but the contagion burned itself out within the landscape before it could spread.

The question that financial stability authorities are now grappling with is not 'how do we prevent another crypto contagion?' It is 'what happens when the interconnections between crypto and traditional finance are substantially larger?' A question that becomes more urgent with every institutional crypto product launch, every bank custody announcement, and every regulatory approval of a spot crypto ETF.

The systemic risk dimensions of the fintech sector are not arguments against fintech. They are arguments for proportionate, forward-looking governance — for regulatory structures designed around systemic importance, and for supervisory investment in the tools needed to monitor genuinely new risks. That governance challenge becomes sharper still when the technology driving fintech's growth is artificial intelligence.

By November 2022, FTX was the second-largest cryptocurrency exchange in the world by trading volume, valued at $32 billion and serving more than 1 million customers. It was headquartered in the Bahamas, under regulatory oversight by the Securities Commission of the Bahamas — a body whose resources were not commensurate with the systemic importance of the institution it supervised.

The collapse, when it came, was rapid. Reporting by CoinDesk revealed that FTX's sister trading firm, Alameda Research, held a significant portion of its assets in FTT — the exchange's own token, whose value was entirely dependent on confidence in FTX itself. When Binance's CEO publicly announced the liquidation of Binance's FTT holdings, confidence collapsed, customer withdrawal requests flooded in, and FTX revealed what subsequent bankruptcy proceedings confirmed: customer deposits had been lent to Alameda Research for trading purposes. The funds were not there.

What the FTX collapse revealed was not primarily a technology failure. It was a governance failure — the failure of a regulatory structure that allowed a company serving over a million customers in regulated jurisdictions to operate under oversight that was inadequate to its scale and complexity. The customers who lost funds had no regulatory protection equivalent to deposit insurance, no resolution regime to return their assets in an orderly manner, and no advance warning from the supervisory system that the institution they were using was not what its marketing materials claimed it was.

The FTX collapse did not produce systemic contagion in the broader financial system because crypto and traditional finance remained, in 2022, substantially compartmentalized. The regulatory lesson — that compartmentalization is a temporary condition and that governance models must be built before the interconnections grow to the point where a crypto crisis becomes a banking crisis — is the most important one this episode provides.

TAKEAWAY: FTX *illustrates the core principle of systemic risk management: the governance infrastructure that prevents catastrophic failures must be built before the failures occur, not assembled in their aftermath. By the time a platform is large enough to be obviously systemically important, the political cost of imposing appropriate governance has become substantially higher than the cost of imposing it earlier would have been.*

CHAPTER NINE

AI and Bank Risk

The Inverted U

*What matters is not whether AI will change the
risk profile of financial institutions. It already has. The
question is whether we understand how — and whether
we can manage the transition before it manages us.*

70%	**€1.9B**	**90%**
of global financial services cloud workloads now run on just three providers— Amazon Web Services, Microsoft Azure, and Google Cloud	*in phantom cash triggered the collapse of the German fintech company Wirecard in 2020*	*of banks report that cyber risk associated with digital financial platforms is now their single largest operational threat*

The chief empirical finding in this book — the one that has the most direct and significant implications for how regulators, bank executives, and policymakers should approach the AI transformation of financial services — was not the one that was expected when the research was designed.

The expectation, going into the analysis of 85 European banks across 21 countries from 2005 to 2022, was that the relationship between fintech AI development and bank risk would be broadly positive: that more AI would mean better risk management, lower default rates, more accurate credit assessment, and ultimately more stable institutions. This is the optimist's thesis, and it has a rigorous academic pedigree. Better information means better decisions. Better decisions mean lower risk. AI provides better information. Therefore, AI should reduce bank risk.

The expectation, alternatively, was that the relationship would be broadly negative: that AI creates new forms of model risk, accelerates competitive dynamics that push banks toward excessive risk-taking, and concentrates systemic vulnerability in ways that outweigh the individual risk management benefits. This is the pessimist's thesis, and it also has a rigorous academic pedigree. Technology that's not fully understood creates risks that are not fully managed. Competitive pressure drives innovation faster than governance. New risks are always harder to see than familiar ones.

Both are correct. That is what makes the finding interesting.

THE INVERTED U: A PRECISE DESCRIPTION OF AN IMPRECISE DANGER

The relationship between fintech AI development and bank risk in the Eurozone data is non-linear. Below a threshold value, increases in fintech AI development are associated with elevated bank risk. Above that threshold, the relationship reverses: further increases in fintech AI development are associated with reductions in bank risk.

The curve is shaped like an inverted U. Risk rises as AI development increases from zero, peaks at the threshold, and then declines as further AI development takes the system past the transition point and into what the research characterizes as the optimization phase.

The economic logic of this shape is not arbitrary. It reflects two overlapping dynamics that operate at different speeds and in different directions.

The first dynamic — risk elevation in the early phase — operates through competitive pressure. When fintech challengers enter a banking market with superior credit models, lower cost structures, and better products, incumbent banks face an immediate and uncomfortable choice: accept declining market share and declining returns or respond by taking on additional risk to maintain performance. The response that regulators most frequently observe is the latter: banks that face significant fintech competition extend credit to borrowers they would previously have declined, expand into less familiar product categories, compress their credit standards under pressure to maintain volumes, and in some cases, increase leverage to offset the margin compression that competition imposes. Each of these responses elevates risk in the short term, and the cumulative effect — across a banking system encountering fintech competition simultaneously — is a period of elevated systemic risk that the linear optimist's narrative doesn't predict.

The second dynamic — risk reduction in the mature phase — operates through the genuine improvements in risk management that mature AI systems deliver. When AI credit models have been trained on sufficient data through sufficient time horizons — including downturns as well as expansions — their predictions become more accurate. When AI fraud detection systems have learned the patterns of fraud at a sufficient scale, their detection rates rise, and their false positive rates fall. When AI operational risk management has been embedded deeply enough in institutional processes, its advantages over rule-based approaches become self-sustaining through the continuous learning that scale provides.

The threshold — the turning point on the inverted U — represents the moment at which the risk management improvements begin to outweigh the competitive dynamics. Banks that have progressed through the early adoption phase have, by definition, survived the competition it creates. They have built AI capability that is beginning to deliver genuine risk reduction. And they have developed the governance structures — model validation processes, human oversight requirements, ongoing monitoring systems — that allow AI to be deployed safely at scale.

THE RESEARCH DESIGN: WHY THE FINDINGS SHOULD BE TRUSTED

The inverted U finding might seem, at first description, like a result that is too neat — a shape that confirms what the researcher wanted to find. This skepticism is healthy. The methodological response to it is the instrumental variable approach that the study employs.

The central challenge in estimating the causal effect of AI on bank risk is that the relationship is potentially endogenous: banks that are already in better shape — better capitalized, better managed, with stronger credit cultures — may both adopt AI more readily and exhibit lower risk scores, making it impossible to determine whether AI is causing the risk reduction or merely correlating with it. Conversely, banks under competitive pressure — higher risk, worse performance — might be both more likely to adopt AI rapidly (as a response to their situation) and to show worse risk metrics initially, creating a spurious positive association between AI adoption and risk elevation that has nothing to do with AI's causal effect.

The instrumental variable approach addresses this by identifying factors that drive AI adoption for reasons unrelated to the bank's underlying risk profile. The study uses two instruments: the interaction between mobile internet penetration in a country and the country's GDP per capita (which captures both the commercial incentives for fintech development and the infrastructure that supports it), and the regulatory environment for fintech in each country (which affects the pace of AI adoption through licensing and compliance requirements). These instruments are correlated with fintech AI development — countries with better mobile infrastructure and more fintech-friendly regulation see faster AI adoption — but are not directly caused by individual bank risk levels.

The instrumental variable estimates confirm the non-linear relationship. The inverted U is not a statistical artefact of endogeneity. It's a causal pattern: AI adoption in earnest causes risk elevation in the early phase and risk reduction in the mature phase, and the instruments provide sufficient variation in the timing of transitions between phases to identify this pattern credibly.

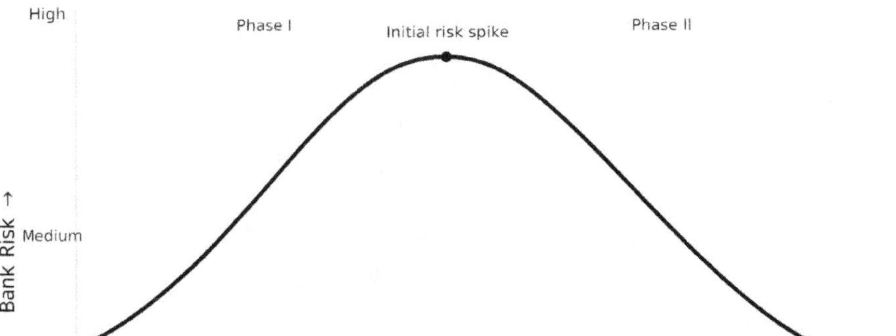

THE THREE-CHANNEL IMF FRAMEWORK

The empirical finding of the inverted U aligns closely with the analytical framework that the International Monetary Fund has developed for understanding AI's systemic risk implications, and the alignment is worth dwelling on because it suggests that the specific Eurozone findings reflect a more general pattern that applies across financial systems at different levels of development.

The IMF's 2023 analysis identifies three channels through which AI affects financial stability.

The first channel is the correlated behavior channel. As financial institutions adopt similar AI architectures — often built on the same foundational models, trained by data scientists who have learned from the same research literature, updated through similar model review cycles — their responses to common market signals become increasingly correlated. When credit conditions deteriorate, and a shared architectural feature causes multiple credit models to simultaneously tighten lending standards, the aggregate credit contraction is larger than the sum of individual institutional responses. The diversity of behavior that historically absorbed macroeconomic shocks — some lenders tightening when others were still expanding, providing a natural buffer — is eroded by the convergence of AI architectures.

The second channel is the opacity channel. Traditional credit models — scorecards, logistic regressions, rule-based systems — have failure modes that are well understood from decades of experience and academic study. Their parameters can be examined, their errors traced, and their biases identified. Machine learning credit models are substantively different in this respect: the relationship between inputs and outputs in a deep neural network is distributed across millions of parameters in ways that resist the kind of direct inspection that allows model failures to be identified before they become significant. When a model learns a misleading pattern—overfitting to a feature in the training data that fails to apply to current market conditions—it can consistently make inaccurate predictions, which may go unnoticed until they result in actual losses.

The EU AI Act's classification of credit scoring as a high-risk AI application, requiring enhanced transparency and explainability, is a direct regulatory response to this opacity concern. But the technical state of explainability in AI is not yet sufficient to fully resolve it: the justifications that current explainability tools provide — feature importance rankings, counterfactual examples, attention weights — approximate rather than fully characterize the decision process of complex models. Regulators who are satisfied with explainability tools that provide plausible narratives for model decisions, rather than genuine causal accounts, may be accepting less assurance than they believe they are receiving.

The third channel is the speed channel. AI systems make decisions — credit approvals, market orders, fraud flags, risk assessments — at speeds that substantially exceed human review capacity. In trading applications, AI operates in microseconds. In credit applications, AI operates in seconds. In fraud detection, AI operates in milliseconds. The speed advantage is, in most cases, commercially beneficial: faster decisions mean better customer experiences and more efficient markets. But it also means that when AI systems make errors, those errors propagate through the financial system faster than any human oversight mechanism can detect and correct them.

The 2010 Flash Crash — in which algorithmic trading systems collectively triggered a trillion-dollar market movement in minutes — is the canonical example. But the financial system has changed substantially since 2010: AI is now embedded not just in trading but in credit origination, risk management, regulatory reporting, and the infrastructure that processes the daily clearing and settlement of trillions of dollars in financial obligations. The speed of exposure has widened, and the governance infrastructure designed to manage it hasn't kept pace.

THE NIST AI RISK MANAGEMENT FRAMEWORK: A PRACTICAL GOVERNANCE TOOL

The National Institute of Standards and Technology's AI Risk Management Framework — published in January 2023 — provides the most operationally useful governance structure for managing AI risk in financial institutions, and its adoption is worth examining in practical detail because it represents the state of the art in AI governance at the institutional level.

The AI RMF organizes risk management around four core functions: Govern, Map, Measure, and Manage.

Govern establishes the organizational structures and accountability mechanisms within which AI risk management occurs. For a financial institution, this means defining who owns AI risk as a category — which executive has ultimate accountability for AI model failures, which board committee reviews AI risk exposure, and how AI risk is integrated into the enterprise risk management framework that governs the institution's overall risk appetite. The Govern function also covers the policies and procedures that define what AI systems can be used for, what human oversight is required for different applications, and what the process is for approving new AI deployments.

The most common failure of Govern-level AI risk management in financial institutions is the classification problem: treating AI as a technology risk rather than a business risk. When AI risk management is housed in the technology function — because AI is perceived primarily as a technical matter — the governance of AI deployment decisions can fall below the seniority level at which the business risk implications are understood. An AI credit model that systematically under-lends to certain demographic groups is not primarily a technology problem. It's a regulatory problem, a reputational problem, and a legal problem — and governing it requires involvement from legal, compliance, risk management, and business leadership, not just from data science and technology.

Map identifies the specific AI risks relevant to each application and context. For a credit decision AI, the Map function would identify the populations at risk of adverse impact, the scenarios under which the model is most likely to fail, the potential for discriminatory outcomes, and the downstream consequences of model error at different rates and in different directions. For a fraud detection AI, the Map function would identify the consequences of false positives — legitimate transactions wrongly blocked — as well as false negatives, and the asymmetry between these error types across different customer populations.

The Map function is where the AI risk framework connects most directly to the compliance and consumer protection obligations that financial institutions already manage. Credit models are subject to fair lending obligations; map-level risk assessment for a credit AI must include an explicit assessment of potential disparate impact across protected demographic categories. Fraud detection systems affect which customers can successfully transact; map-level assessment must include an evaluation of which customer populations are disproportionately affected by false positives, and whether that disproportionality reflects or amplifies existing financial exclusion.

Measure develops the metrics and assessment methods that allow AI risks to be tracked quantitatively over time. For AI in financial services, the critical measurements include model performance on hold-out samples and in production, performance drift over time, performance differentials across demographic groups, adversarial robustness, and the frequency and nature of model failures and overrides. The measurement discipline also covers ongoing monitoring: systems for detecting when model performance is degrading before the degradation becomes important, and mechanisms for escalating concerning patterns to appropriate human oversight.

Manage puts in place the controls and processes that reduce AI risks to acceptable levels and respond to failures when they occur. This includes human-in-the-loop requirements for high-stakes decisions, model override mechanisms that allow trained humans to countermand AI recommendations, incident response procedures for model failures, and continuous improvement processes that incorporate the lessons of production experience into model updates.

THE EU AI ACT'S HIGH-RISK CLASSIFICATION FOR FINANCE

The EU AI Act's categorization of several financial AI applications as high-risk — subject to the most stringent transparency, oversight, and documentation requirements in the regulation — has significant practical consequences for European financial institutions and, through the extraterritorial application that characterizes EU regulation, for non-European institutions serving EU customers.

The financial applications designated as high-risk include AI used in credit scoring, insurance pricing, claims assessment, and certain fraud detection systems. For each of these, the AI Act requires: conformity assessment before deployment, documenting that the system meets the regulation's technical requirements; a register of high-risk AI systems, maintained by the deploying institution and accessible to regulators; ongoing monitoring of system performance in deployment; a mechanism for human oversight and the ability to override automated decisions; transparency to users about when AI is being used to make high-stakes decisions about them; and explainability of individual decisions at the request of the affected person.

The explainability requirement is the most operationally demanding. A person whose credit was declined by an AI model has, under the AI Act, the right to receive an explanation of that decision that they can meaningfully engage with. The decision can't be characterized solely as 'the algorithm declined your application' — the institution must be able to identify the factors that drove the decision and explain them in terms the applicant can understand and potentially contest.

The practical implementation of this requirement has revealed a tension in the AI Act's architecture: the most accurate credit models — deep neural networks with many layers and millions of parameters — are also the least explainable in the sense the regulation requires. The models that are most explainable — linear scorecards, shallow decision trees — are less accurate. The financial institution deploying AI for credit decisions must choose between accuracy that it can't fully explain and explainability that comes at a cost in accuracy. The regulation's answer to this tension is that explainability is a non-negotiable requirement — which means, in effect, that the opaquest models can't be used for high-risk decisions, regardless of their accuracy advantage.

This isn't obviously wrong as a policy choice. The right to understand the algorithmic decisions that affect your financial life is a legitimate principle. But it has consequences — including the possibility that financial institutions in jurisdictions without equivalent requirements will deploy more accurate but less explainable models, capturing credit risk management advantages that EU institutions can't access. Whether this trade-off is acceptable is ultimately a values question, not a technical one.

HSBC and Quantexa: Network Intelligence Against Financial Crime

HSBC's deployment of AI-powered network analytics through its Quantexa partnership represents the mature phase of AI risk management in practice. The core insight behind the partnership is that fraud and money laundering are almost always network phenomena — involving multiple entities, accounts, and transactions that individually appear innocuous but collectively reveal criminal patterns. Rules-based monitoring systems evaluating transactions in isolation miss these patterns entirely.

Quantexa's analytics layer maps the relationships between entities — individuals, companies, accounts, addresses — across HSBC's entire transaction history, identifying unusual network configurations associated with criminal activity. The system learns from confirmed cases, continuously improving its pattern recognition.

The upshot is a dramatic reduction in both false negatives — fraudulent transactions that slip through — and false positives — legitimate transactions wrongly flagged — reducing both direct fraud losses and the enormous compliance costs associated with investigating false alarms. HSBC has reported significant improvements in detection accuracy and reductions in false positives, lowering investigation costs.

This outcome is exactly what the inverted U predicts for the mature phase of AI adoption: the competitive pressure that elevated risk in the early phase has been absorbed, the models have matured past their initial instability, and the genuine risk management advantages of AI — more accurate pattern recognition at greater scale — are delivering measurable stability benefits.

TAKEAWAY: *HSBC's experience with network analytics illustrates the optimization phase of the inverted U: when AI systems have been refined through sufficient operational experience and trained on sufficient data, they deliver risk management outcomes that no rules-based approach can match.*

Klarna, the Swedish buy-now-pay-later company, makes credit decisions in approximately 0.3 seconds, assessing more than 200 variables — including behavioral signals from the customer's current browsing session — to determine creditworthiness at the point of checkout.

The platform processes hundreds of millions of transactions annually, generating a training dataset for its credit models that no traditional lender can replicate. Each decision improves the model. Each model improvement reduces default rates. Reduced default rates improve the economics of the business — and, according to the research on the inverted U, improve the stability of the broader financial system in which Klarna participates.

Klarna reports lower loss rates on comparable short-term purchases than traditional credit card lending. This is the mature-phase AI advantage: more information, more accurately processed, producing better decisions for both lenders and borrowers.

The governance question Klarna raises is equally important: a company making credit decisions at this speed and scale, about this many borrowers, must have model governance infrastructure that matches the speed and scale of its decisions. The explainability requirements of the EU AI Act — which classifies BNPL credit assessment as high-risk AI — will require Klarna to provide comprehensible accounts of individual decisions to customers who request them. Building that capability, for models operating at 0.3 seconds per decision, is one of the defining technical challenges of AI governance in financial services.

TAKEAWAY: *Klarna illustrates that the optimization phase of the inverted U is commercially achievable — but that achieving it doesn't eliminate the governance challenges of operating AI at scale. Model accuracy and model accountability are both necessary; neither is sufficient without the other.*

The NIST AI Risk Management Framework: Practical Application to Financial Services

The National Institute of Standards and Technology's AI RMF — Govern, Map, Measure, Manage — provides the most operationally complete governance architecture currently available for financial institutions deploying AI at scale. Its four functions map directly onto the regulatory requirements of the EU AI Act's high-risk AI provisions and the FSB's principles for AI governance in financial services.

Govern establishes accountability: which executive owns AI risk, which committee reviews AI model deployments, and how AI risk integrates into enterprise risk management. The most common failure at this level is treating AI as a technology risk rather than a business risk — housing AI governance in the IT function rather than at the level where business risk consequences are understood.

Map identifies application-specific risks: for a credit AI, which populations are at risk of adverse impact, what failure modes are most likely, and what discrimination potential exists. For a fraud detection AI: what false positive rate is acceptable; which customer populations are disproportionately affected by blocking; how is the model's performance monitored for distributional shift.

Measure and Manage complete the cycle: metrics that track model performance in production, governance structures that enable human override of AI decisions in high-stakes contexts, and incident response procedures that address model failures when they occur. Together, these four functions constitute the governance infrastructure that regulators in every major jurisdiction are increasingly requiring — and that institutions without it are increasingly vulnerable to enforcement action.

Source: NIST AI Risk Management Framework (AI RMF 1.0), January 2023; European Banking Authority, Guidelines on internal governance (EBA/ GL/2021/05).

CHAPTER TEN

Regulation and Arbitrage

Racing the Referee

Regulators are always fighting the last war. But the soldiers of finance are already engaged in the next one. The question is whether the gap between them is manageable or catastrophic.

70+	300%+	$6B	45%
countries operating dedicated fintech regulatory sandboxes	increase in global fintech regulation enacted since 2018	global fintech regulatory compliance spend, 2025	of fintech firms citing regulatory uncertainty as their primary barrier to growth

When I was studying for my MSc in Global Central Banking and Financial Regulation with Warwick Business School in collaboration with the Bank of England, I found myself in a space that felt very different from anything I had experienced before. It was one of the first times I could genuinely engage with central bankers, academics, and practitioners from across the world in one shared environment. The conversations were not theoretical exercises. They were alive, sometimes tense, often revealing. What stood out to me very early on was not just the diversity of views, but how far apart those views could be, even when everyone was discussing the same issue.

The online forums made this even clearer. We would debate topics like central bank digital currencies, crypto assets, and systemic risk. On the surface, everyone was asking similar questions. Underneath, the answers came from entirely different worlds. Regulators approached these issues with a deep sense of responsibility. Their instinct was to protect the system, to guard against instability, to avoid outcomes that could harm the public. When Bitcoin came up, the tone from regulatory voices was firm. It was seen as volatile, unpredictable, and potentially dangerous in the wrong hands. There was a clear concern that something so unstable should not be presented to ordinary people as a viable financial instrument.

Academics approached the same question from a different angle. The discussion shifted to whether Bitcoin could realistically function as money. Could it serve as a unit of account, a medium of exchange, or a store of value? The arguments were structured, grounded in theory, and often led to the conclusion that Bitcoin, in its current form, could not meet these criteria in a meaningful or sustainable way. The limitations were clear. Price instability made it difficult to anchor value. Transaction costs and speed posed practical challenges. The idea of it replacing traditional money did not hold up under closer scrutiny.

Practitioners saw something else entirely. For many of them, Bitcoin was not just an asset or a flawed currency. It represented a shift in how financial systems could be built. It was about decentralization, about removing intermediaries, about creating new forms of financial access that did not rely on traditional institutions. There was an energy in those discussions that was hard to ignore. Even when the limitations were acknowledged, the sense of possibility remained strong. It was not about what Bitcoin was at that moment. It was about what it could become.

Sitting in the middle of these conversations, I began to understand why that program mattered so much. It forced these perspectives into the same room. It did not resolve the differences, but it exposed them in a way that made them impossible to ignore. The gap between regulators, academics, and practitioners was not a minor misalignment. It was structural. Each group was operating with a different set of assumptions, a different time horizon, and a different definition of risk.

That realization has stayed with me. It explains a lot about how financial regulation has evolved over the past decade. It also explains why arbitrage continues to exist in the form that it does. Arbitrage is often described in technical terms, but at its core, it is about gaps. It is about the space between how something is understood and how it is governed. When those spaces open up, they create opportunities. Some of those opportunities are harmless. Others are not.

In 1987, Alan Greenspan became chairman of the Federal Reserve. Two months into his tenure, the stock market crashed 22% in a single day — still the largest one-day percentage decline in US market history. Greenspan's response — providing liquidity to the financial system, signaling that the Fed stood behind the markets — is credited with preventing the crash from becoming a depression. He served for nearly two decades, presided over the longest peacetime economic expansion in US history, and was celebrated, near the end of his tenure, as the greatest central banker of his generation.

In 2008, the financial system he had helped build collapsed in ways that his framework did not anticipate, and his tools were inadequate to prevent. The instruments that had been developed to manage the risks of the 20th-century financial system — leverage limits adjusted to balance sheet assets, capital requirements tailored to loan categories, supervisory examination cycles tailored to annual reporting — were designed for a financial system that processed its riskiest decisions over months and years. By 2008, the financial system was processing its riskiest decisions in seconds, through instruments whose risk characteristics had never been fully mapped, over networks of interconnection whose topology no regulator had systematically documented.

This isn't a story about Greenspan's failures. It's a story about the structural lag that characterizes the relationship between financial innovation and financial regulation in every era, and that poses the central challenge of financial governance in the age of AI.

Every major financial innovation creates risks that the existing regulatory framework was not designed to manage. The regulatory response — learning from experience, often from crisis, what the new risks are and how to govern them — always comes after the innovation rather than before it. The question is not whether this lag exists. It always does. The question is how wide it grows before the regulatory system catches up, and whether the consequences of that gap are manageable or catastrophic.

In the AI era, the lag is structured differently from any previous period in financial regulation history. Previous innovations — derivatives, securitization, algorithmic trading — were comprehensible to the regulatory apparatus that had to govern them, even if the specific risks they created were not fully understood at first. Derivatives were complex, but they were contracts between identifiable counterparties with identifiable exposures. Securitization was opaque, but the underlying assets were mortgages and auto loans with recoverable risk histories. The challenge was primarily one of information: regulators needed more and better information about the risks being taken. The analytical framework was adequate. The data was missing.

AI presents a special challenge. The systems being built are not just opaque in the sense that they are inadequately documented. They are opaque in the deeper sense that their internal workings — the learned representations, the weights, the attention patterns that determine their outputs — can't be directly inspected and interpreted even in principle, with the analytical tools currently available. A senior credit officer at a bank can explain why a specific loan was declined. The AI model that made the same decision can't be interrogated to yield an equally clear account. The opacity is intrinsic to the architecture, not a documentation failure.

This is why the regulatory challenge of AI is different in kind from the challenges of previous financial innovations, and why addressing it requires not just more of what existing regulatory structures do — more information, stricter standards, bigger penalties — but a truly new approach to the question of what financial regulation is and what it can accomplish.

THE REGTECH IMPERATIVE: SUPERVISING AI WITH AI

I remember reading an article while I was in Shanghai that described how difficult it would be for regulators to fully analyze Ant Group's data after the decision to halt its IPO. The issue was not just scale. It was the nature of the data itself. The volume was immense, but more importantly, it was dynamic, behavioral, and constantly evolving. Traditional regulatory tools were not designed to process that kind of information in real time. The system had outgrown the tools that were meant to oversee it.

Regulation has always followed innovation, but the distance between the two is widening. In earlier financial eras, that gap was manageable. The system moved at a pace that allowed regulators to catch up, to adapt, to build frameworks that could respond to new risks. What we are seeing now feels different. The speed is higher. The complexity is greater. The feedback loops are tighter. The system does not wait. This is where my concern lies. Artificial intelligence will not simply add another layer to this dynamic. It will change the structure of the gap itself. There is a possibility that AI could help narrow the distance between regulators and the systems they are trying to oversee. With the right tools, regulators could process information faster, identify risks earlier, and respond with greater precision.

There is another possibility that is harder to ignore. AI could deepen the divide. If advanced systems are developed and deployed primarily by large financial institutions and technology platforms, the informational advantage they already hold will expand even further. The asymmetry between those building the systems and those regulating them could reach a point where oversight becomes reactive by default. If that happens, the consequences will not be abstract and the gap will keep widening, making it impossible for regulators to catch up. Regulatory frameworks that were designed for a different kind of financial system will struggle to remain relevant. They were built around institutions, products, and risks that are easier to observe and classify. AI-driven systems do not fit neatly into those categories. They operate across boundaries. They learn. They adapt. They make decisions in ways that are not always transparent, even to the people who design them.

This is why regtech cannot remain static. It must evolve in a way that matches the systems it is meant to oversee. That does not mean simply digitizing existing regulatory processes. It means rethinking how regulation works at a fundamental level. AI needs to be part of that process, not as an afterthought, but as a core component. There is a practical dimension to this. Data quality matters. Access matters. Interpretation matters. If regulatory systems are trained on incomplete or outdated information, they will produce conclusions that do not align with reality. The risk is not just that they miss problems. It is that they create a false sense of control.

It is becoming clearer to policy makers that you cannot supervise AI-powered finance with non-AI-powered supervision. The gap between the analytical capability of the AI systems operating in financial markets and those of the human supervisors overseeing those markets is widening with every new model deployed, every new product launched, and every new risk dimension created. Closing that gap — or at least preventing it from widening to catastrophic proportions — requires investing in AI-powered supervisory tools with the same urgency and seriousness that the financial institutions being supervised are investing in AI-powered financial tools.

This isn't a technology advocacy statement. It is a basic observation about the mathematics of oversight. When the system being supervised is capable of processing millions of credit decisions per day, a supervisory framework whose primary tool is annual examination of sampled transactions cannot catch problems before they become systemic. When the product being reviewed is generated by a model with hundreds of millions of parameters, a supervisory assessment framework based on manual document review cannot evaluate the model's risk properties. The mismatch is not a matter of effort or resources at the margin. It is a structural gap that only technology can close.

The Bank of England and the Financial Conduct Authority's joint innovation lab — the FCA Sandbox's technology stream — has been the most visible public investment in AI-powered supervision by any major financial regulator. The BoE and FCA have deployed machine learning systems to monitor social media, news feeds, and financial market data for early warning signals of emerging financial stress, to analyze the text of regulatory filings for patterns associated with future compliance failures, and to process the transaction reporting data that the European Market Infrastructure Regulation requires financial institutions to submit, identifying anomalies that would take human analysts weeks to find.

The results are encouraging. The FCA's NLP-based monitoring system identifies potential mis-selling campaigns from financial promotions data weeks before customer complaints would reveal the same issue. The BoE's market intelligence AI identifies unusual patterns in options pricing and credit default swap spreads that historically precede financial stress events. Neither system is perfect — both generate false positives that require human assessment and miss some events that would later be identified as significant. But both represent a genuine improvement in supervisory capability over the human-only alternative, and both are continuing to improve as their training datasets grow and their architectures are refined.

The global picture of regtech investment is more uneven. European regulators — the ECB's supervisory arm, the EBA, and several national supervisors — have made substantial investments in advanced analytics and AI-powered surveillance. The Federal Reserve and the OCC in the United States have built meaningful capabilities. Most developing-country regulators remain in the early stages of regtech deployment, precisely in the markets where the fintech adoption is fastest, and the supervisory challenge is greatest.

This asymmetry — advanced AI-powered supervision in markets that also have more mature regulatory frameworks and better-capitalized institutions, while developing markets face rapid fintech adoption without equivalent supervisory capability — is itself a systemic risk. It creates the conditions in which the most harmful outcomes of poorly governed fintechs are most likely to occur: in markets where regulatory capacity is weakest, and the populations most dependent on fintech for financial access are most vulnerable.

THE EU'S COMPREHENSIVE REGULATORY ARCHITECTURE

The European Union has produced the most ambitious attempt in the world to build a coherent regulatory structure for AI-powered digital finance — one that addresses multiple risk dimensions simultaneously through a set of interconnected regulations that, taken together, create something approaching a full governance model.

MiCA — Markets in Crypto-Assets — provides the world's first broad framework for crypto assets, including stablecoins. It establishes authorization requirements, capital and reserve requirements, consumer protection rules, and market integrity standards that apply uniformly across the 27 EU member states. By closing the regulatory arbitrage that had previously allowed crypto businesses to operate in the EU from minimally regulated offshore jurisdictions, MiCA creates the conditions for legitimate crypto businesses to operate safely while substantially reducing the harm potential of fraudulent or inadequately capitalized operators.

DORA — the Digital Operational Resilience Act — addresses cloud concentration and operational resilience. Its most significant conceptual innovation is the extension of financial regulatory oversight to critical third-party technology providers — cloud companies, data centers, AI model vendors — that financial institutions depend on but that were previously outside the regulatory perimeter. DORA requires financial institutions to map their critical technology dependencies, assess vendor resilience, maintain contingency plans for vendor failure, and participate in system-wide resilience testing that includes scenarios in which critical shared infrastructure fails.

The AI Act classifies financial AI applications by risk level and imposes proportionate requirements. Credit scoring, insurance pricing, and certain fraud detection applications are classified as high-risk, requiring conformity assessment, human oversight mechanisms, explainability capabilities, and registration in a public database. This classification system — risk-based and technology-neutral — is the right framework for AI regulation in finance, because it focuses regulatory attention on the applications where errors cause the most harm rather than on the technology itself.

PSD3 — the revised Payment Services Directive — updates the open banking framework to address emerging challenges in the evolving payments landscape, including better protection against payment fraud, stronger requirements for customer authentication, and improved rules for the management of payment data.

The Digital Markets Act addresses platform competition — ensuring that the largest digital platforms cannot use their market power to foreclose competition in adjacent financial services markets. This provision is particularly relevant to the Techfin challenge identified in Chapter One: the risk that platform companies use their control of the interface between consumers and financial services to extract rents from both consumers and financial service providers, without being subject to the competitive disciplines that apply in markets with multiple competing platforms.

Taken together, these five regulations constitute the broadest framework in existence for the governance of digital finance. Their limitations are practical rather than conceptual: the regulatory capacity to enforce them, the technical expertise to assess compliance with them, and the international coordination required to prevent regulatory arbitrage around them remain substantially incomplete.

THE ARBITRAGE PROBLEM AND ITS PARTIAL SOLUTIONS

The structural challenge of fintech regulation has not changed since the first fintech regulation was written: financial services are global; regulation is national. As long as meaningful differences in standards exist between jurisdictions, commercial incentives will push activity toward the most permissive available territory. The history of regulatory arbitrage in financial services is long, and its consequences are largely one of harm to consumers and systems in the least-regulated markets.

Digital finance has made this problem worse in specific ways. Geographic arbitrage — locating operations in the most permissive jurisdiction while serving customers globally — has always been possible but was constrained by the practical costs of maintaining operations in remote locations. Digital finance eliminates most of those practical costs. An exchange can be domiciled in the Seychelles and serve customers in 180 countries through nothing more than a website and a payment processor. The regulatory arbitrage is available at effectively zero cost of physical relocation.

The FTX collapse illustrated this in the starkest possible terms. Headquartered in the Bahamas, operating under Bahamian regulatory oversight that was simultaneously limited by the regulator's own capacity and by the company's active management of its supervisory relationships, FTX served institutional and retail customers across major regulated markets, including the United States, Europe, and Asia — markets whose regulators had no direct supervisory authority over the Bahamian entity. When the fraud was uncovered, those customers had no recourse to the regulatory protections that they would have had if a regulated domestic entity had defrauded them.

The solutions to regulatory arbitrage are partial and demanding. Mutual recognition agreements — in which jurisdictions agree to recognize each other's regulatory standards as equivalent — reduce arbitrage incentives by expanding the territory within which a single regulatory passport is valid. The EU's equivalence regime and the UK's post-Brexit framework for managing equivalence relationships are examples of this approach. But they work best between jurisdictions with genuinely comparable regulatory standards, and they create their own risks: a jurisdiction deemed 'equivalent' that subsequently lowers its standards can export regulatory failures into the markets that have granted it recognition.

Extraterritorial regulation — applying one jurisdiction's rules to activities that occur outside its borders but affect its consumers or financial system — is the other major tool. The EU's General Data Protection Regulation has been the most significant exercise of extraterritorial digital regulation: it applies to any company that processes the data of EU residents, regardless of where that company is established. MiCA takes a similar approach for crypto assets: it applies to crypto businesses that offer services to EU customers, regardless of where those businesses are located.

Extraterritorial regulation is effective against large companies with EU market exposure significant enough to justify compliance investment. It is less effective against smaller operators targeting EU consumers from offshore, where the calculus of potential EU enforcement action against the expected profits of non-compliance tilts toward non-compliance. And it creates diplomatic friction when the regulated activities are legitimate in the jurisdictions where they occur, making multilateral coordination essential for its long-term sustainability.

The direction of travel is clear: more international coordination on AI governance standards, more mutual recognition of equivalent regulatory frameworks, and broader extraterritorial application of the standards developed by the jurisdictions with the most sophisticated regulatory frameworks. The pace is not clear. History suggests that coordination of this kind occurs more quickly after crises than before them.

Policy Framework — Muganyi et al. (2022): Ten Principles for Balanced Fintech Growth

The China financial development research concludes with a policy framework built on ten priorities that apply across markets at different stages of fintech development: (1) develop fintech-specific regulation rather than extending bank regulation wholesale; (2) invest in regtech supervisory infrastructure to match the analytical power of supervised entities; (3) create regulatory sandboxes that test products at scale before authorizing market-wide deployment; (4) establish data governance that protects consumers while enabling innovation; (5) treat interest rate liberalization as a complement to fintech, not a substitute for it.

The remaining five priorities address systemic and international dimensions: (6) pursue international coordination to prevent regulatory arbitrage — a recommendation that has become more urgent, not less, as the global interconnection of digital financial markets has deepened; (7) build bank-fintech partnership frameworks that maintain systemic oversight while enabling innovation; (8) invest in financial literacy as a complement to financial access; (9) develop resolution frameworks for systemically important fintech platforms before they become necessary; (10) create dedicated research institutions on fintech development with international cooperation mandates.

These ten principles are not abstract recommendations. Each emerges directly from an empirical observation in the research: that regulatory quality is a complement to fintech development rather than a constraint on it, that the development dividend of fintech is larger in better-governed markets, and that the systemic risks of poorly governed fintech are real, growing, and addressable through proportionate regulation designed for the technology rather than adapted from frameworks built for its predecessors.

Source: Muganyi, T., Yan, L., Yin, Y. et al. (2022). Financial Innovation 8(29); analysis extended with additional panel data.

CHAPTER ELEVEN

Fintech Models of the Future

What Comes Next

The future is already here — it is just not evenly distributed yet. The question isn't what fintech will become, but who will benefit from what it becomes, and on whose terms.

65%	**$7T**	**2.1B**
of financial transactions expected to be AI-mediated by 2030	*embedded finance market projected by 2030 (Bain & Company)*	*Africa's population by 2040 — the continent that will most determine fintech's next chapter*

My first job after college was with a state-owned bank. It was not the kind of place where you talked about innovation or disruption. It was a place where you learned how the system actually worked. During my training, I rotated across departments to understand the bank from the inside. Operations, compliance, retail banking. I spent time in all of them, but I found myself drawn to credit and risk. That was where decisions were made that could change someone's life. Most of those decisions started with a file. A simple A4 folder with pay slips, identity documents, and a few standard forms. Sometimes there was a guarantor attached. It looked complete. It looked official. It felt like enough.

It wasn't.

We had to visit the clients to understand what was really going on. You would stand on a farm or sit in a small business and realize very quickly that the file told only a fraction of the story. Cash flow moved in ways that could not be captured neatly. Income came in waves. Costs appeared without warning. Decisions were shaped by things no document could explain. Even with those visits, most of what we did still depended on the structure we had built. The metrics were fixed. The thresholds were clear. If someone fit within them, the process moved forward. If they didn't, it stopped. It was not personal. It was how risk was managed.

Looking back, I began to notice a pattern. Lending did not flow evenly. It moved toward those who already looked stable on paper. Those who had formal income, predictable records, and a history the system could understand. Everyone else remained outside, not because they lacked potential, but because they did not fit the model. Sometimes it was also about who you knew and how those social networks interacted.

At the time, I accepted that as the nature of banking.

Now, I am less certain.

Working as a consultant today, I see how much the tools have changed. Banks now hold detailed transaction histories, behavioral data, and digital footprints that go far beyond anything we had access to back then. The visibility is deeper. The speed is faster. The scale is far greater. And yet, something about it feels familiar. The data may be richer, but it is still organized around predefined fields and metrics. It still tries to describe people in ways that the system already understands. It still reduces complexity into something that can be processed, scored, and approved or rejected. In many ways, it feels like a more sophisticated version of the same file I used to work with. That is where the question begins to take shape for me.

If we have all this data, why does banking still feel so rigid? Why does it still struggle to adapt to the realities people live in? Why does it still favor those who already fit within its structure? The answer lies in how the models are built. They are designed to measure what is already known. They are not designed to explore what is possible. This is where the future of fintech starts to look different.

Artificial intelligence creates an opportunity to move beyond static representations of people and businesses. It allows for something more dynamic, more responsive, closer to how the real world behaves. Instead of relying on past records alone, systems can begin to understand patterns as they unfold. They can adapt. They can learn. I think about the farms we used to visit. What if, instead of relying on a snapshot of income and expenses, we could simulate how that farm would perform under different conditions? What if we could see how changes in rainfall, input costs, or market prices would affect output before a loan is ever disbursed? What if the business owner could do the same? What if decisions were not based on a static view of the past, but on a living model of the present? That is not a distant idea anymore. The building blocks are already here. Data is available. Models are improving. The ability to simulate, predict, and adapt is no longer theoretical. What is changing is not just the technology, but the way financial systems can be designed. Banking does not have to remain visible in the way it has always been. It does not have to feel like a process you go through. It can become something that works in the background, adjusting to your needs, responding to your situation, supporting decisions as they happen. It can become more human, not by removing structure, but by making that structure more flexible.

The models of the future will not start with forms. They will start with context.

They will understand that income is not always monthly, that risk is not always static, and that potential cannot always be captured in a number. They will respond to behavior in real time. They will learn from outcomes. They will adjust as conditions change. This does not remove risk. It changes how we understand it. The role of the bank begins to shift. It is no longer only about evaluating what has already happened. It becomes about engaging with what is happening now and what could happen next.

That changes the relationship between institutions and the people they serve. It also raises new questions. Who builds these models? Who has access to them? Who understands how they work? If the answers to those questions are concentrated, then the advantages they create will be concentrated as well. The gap between those inside the system and those outside it could widen in new ways. That is something we cannot ignore. The future of fintech is not just about capability. It is about direction.

It is about whether these systems are designed to expand participation or reinforce existing boundaries. It is about whether they make finance more accessible in a meaningful way, or simply more efficient for those already included. I often think back to those early days in the bank, sitting with files that tried to tell a complete story but never quite could. What we were really doing was making decisions with limited visibility and a strong need for certainty. Today, we have the chance to move beyond that.

We can build systems that see more, understand more, and respond more effectively. But that will not happen automatically.

In the developing world a number of banks are deploying Large Language Model-enabled chatbots to help their customer service representatives handle complex queries from small business clients. The deployments are not always announced or celebrated as a breakthrough. It is simply a practical decision by practical institutions: the AI can process more queries, more accurately, more quickly, and at lower cost than the human-only approach it supplements. The technology has become cheap enough, good enough, and accessible enough that a bank anywhere in the world can deploy it as a routine operational tool.

This is what the AI inflection point actually looks like in practice. Not the dramatic headline deployment of a cutting-edge system by a leading institution in a major market. It is the quiet adoption by institutions in markets that previously could not have accessed the technology, making decisions that would have required expensive human expertise now possible at a cost the institution can sustain. The inflection point is not a single moment. It is a threshold that different institutions in different markets cross at different times, with different consequences — and the aggregate effect of those crossings is the transformation that the coming decade will be defined by.

What comes next is partly visible and partly not. Technology trajectories are clearer than governance outcomes because technology development is more predictable than the political and institutional choices that shape how technology is deployed and who benefits from its deployment.

This chapter attempts a considered account of both.

THE ARCHITECTURE OF WHAT'S COMING

Three technological developments, each already underway and each accelerating, will define the next phase of AI in financial services.

The first is the transition from AI as a tool that assists human decisions to AI as an agent that makes decisions autonomously. The current generation of AI applications in finance — credit scoring models, fraud detection systems, robo-advisory platforms, customer service chatbots — are, in the language of AI research, narrow AI: systems trained to perform specific tasks within defined parameters, producing outputs that humans then use as inputs to their own decisions. The emerging generation — variously described as agentic AI, autonomous AI, or AI with agency — will go further: managing financial positions, negotiating financial contracts, executing compliance processes, and making financial decisions on behalf of users with minimal moment-to-moment human oversight.

The financial services implications of this transition are profound. An AI agent that monitors a client's financial position continuously, rebalances their portfolio when market conditions warrant, renews their insurance at the optimal time, moves their savings into higher-yield accounts when rate differentials justify it, and alerts them to financial risks before those risks materialize — that agent is performing the functions that a full-service private banker performs for wealthy clients, at a cost that makes the equivalent service available to anyone with a smartphone.

For financial inclusion, this is potentially far-reaching. The gap between the financial management available to a wealthy client of a sophisticated private bank and the financial management available to a low-income customer of a basic bank account is, in large measure, a gap in the human advisory resources that can be applied to each relationship. An AI agent that closes this gap — not by providing worse service to wealthy clients but by providing comparable intelligence to clients of all income levels — could represent the most significant equalization of financial capability in history.

The governance challenge is equally significant. An AI agent that acts on behalf of a user — taking financial decisions without the user's specific authorization for each action — must be governed by frameworks that establish accountability for its actions, mechanisms for users to understand and override its decisions, and protections against the forms of manipulation that autonomous financial agents create new opportunities for. The regulatory infrastructure for agentic finance is in its early stages. The technology is not.

The second development is the deepening of embedded finance — the integration of financial services into non-financial products, platforms, and experiences. The first generation of embedded finance — buy-now-pay-later at checkout, insurance at booking, credit at point of sale — was a relatively plain extension of existing products into new distribution channels. The AI-powered next generation will be qualitatively different: financial products that are designed specifically for the context in which they appear, dynamically priced for the specific customer at the specific moment, and delivered as a seamless component of an experience rather than as an add-on.

When an agricultural input platform knows, from satellite imagery and weather data, that a specific farmer's region is in the third week of a drought, and the platform embeds an index insurance payout into the farmer's account without any application process or claim submission, it has performed something that no traditional insurance company has ever been able to do: identify a need, assess its validity, and deliver a financial product in the moment it is needed, before the farmer is aware that they need to act. This is embedded finance at its most powerful — and it is the model that AI enables at scale.

The third development is the emergence of African and South Asian AI financial infrastructure as a global force rather than a local solution. The combination of demographic scale — Africa will have the world's largest working-age population by 2035 — and the leapfrog dynamic that characterizes markets building financial infrastructure from scratch in the AI era creates conditions for innovation that markets encumbered by legacy systems cannot replicate. Some of the most interesting AI financial tools set for deployment in 2026 were built not in Silicon Valley or London but in Lagos, Nairobi, Accra, and Bangalore — by teams solving problems that incumbents in mature markets have no experience with, using AI tools that are now globally accessible.

WHAT THE FUTURE OF MONEY LOOKS LIKE

The monetary order of 2040 will look substantially different from that of 2026, and the outlines of what is coming are already visible in the technology deployments and regulatory frameworks taking shape today.

The most likely structural change is the emergence of a layered monetary system, in which multiple forms of digital money coexist and interoperate: central bank digital currencies providing the state-backed, privacy-protected foundation; regulated stablecoins providing cross-border payment efficiency and access to dollar-denominated value in markets where local currency instability creates demand; and commercial bank deposits continuing to provide the credit intermediation function that neither CBDCs nor stablecoins currently serve.

In this layered system, the relationship between money and data will be different in kind from what exists today. Every monetary transaction will generate a digital record. Much of that record will be held by private parties — the platforms, the payment processors, the banks — but the governance models being established now, from the EU AI Act to India's Account Aggregator framework to the evolving CBDC privacy architectures, will shape how that data can be used, who can access it, and on whose terms.

The development implications of this architecture are, in aggregate, positive — provided the governance choices that are being made now reflect the interests of the people who are currently most excluded from the formal financial system, rather than the interests of the financial institutions and technology platforms that are building it.

This isn't guaranteed. The history of financial system design is a history of systems designed primarily by and for those who were already well-served — and later extended, through political pressure, regulatory mandate, and occasionally commercial discovery of underserved markets, to include those who were initially left out. The fintech era has been more inclusive in its design intent than any previous wave of financial innovation. The AI era needs to be more deliberate still.

THE RISK HORIZON: WHAT COULD GO WRONG

An honest account of what comes next must acknowledge that the pathway from the current state to the positive outcome described above is not guaranteed, and that several foreseeable developments could push the arc in a worse direction.

The first risk is algorithmic discrimination at scale. AI credit models trained on historical financial data encode the lending patterns of historical financial systems — patterns that, in most major economies, were explicitly or implicitly discriminatory along racial, gender, and geographic lines. A model trained on historical credit decisions in the United States will learn that Black applicants default more often — not because race is causally related to credit-worthiness, but because Black applicants have historically had access to inferior financial services, lived in communities with less economic opportunity, and been charged higher rates for credit, all of which affected their ability to repay. A model that learns this pattern will perpetuate it, declining or pricing out Black applicants at rates that replicate historical discrimination while appearing objective because the decision came from an algorithm rather than a human.

The technical tools to detect and correct this bias exist. The regulatory requirements mandating their use are being developed, most notably in the EU AI Act's fairness requirements for high-risk AI systems. But the implementation gap — between what the tools can do and what deployed systems actually do — remains large, particularly in markets outside the EU where fairness requirements are less well developed.

The second risk is the concentration of AI capability in a small number of very large platforms. AI at scale requires three inputs that exhibit strong economies of scale: data, compute, and research talent. All three are concentrated among companies with the largest data assets, the most powerful computing infra-structure, and the greatest ability to recruit and retain the best AI researchers. The financial services industry is not insulated from this concentration: the largest banks and the largest technology companies have the most to gain from the AI transition, because they have the resources to invest in AI capability that smaller institutions cannot match.

If the AI transition in financial services concentrates market power further in the institutions that are already dominant — and the early evidence suggests that this is the direction of travel absent deliberate policy intervention — the result could be a financial system that is more efficient in aggregate but less competitive, less diverse, and less responsive to the needs of populations that the dominant institutions have historically under-served.

The third risk is the widening of the global AI divide. The countries with the most sophisticated AI capabilities — the United States, China, the United Kingdom, and a handful of others — are building AI financial infrastructure that reflects their own market structures, regulatory priorities, and cultural assumptions. When this infrastructure is exported to lower-income markets — as it increasingly is through the global platforms that deploy the same systems across multiple markets — it brings assumptions that may not translate well. A credit model trained on US consumer data will perform poorly on Kenyan borrowers with different income patterns, different collateral norms, and different risk characteristics. An AI customer service system trained on English-language financial interactions won't serve Tamil or Swahili, or Amharic speakers with equivalent quality.

Addressing the global AI divide requires deliberate investment in AI development that reflects the markets it will serve — which means supporting African and South Asian AI researchers, building training datasets that reflect the diversity of global financial experience, and designing regulatory frameworks that incentivize AI deployment in underserved markets rather than exclusively in the profitable ones.

THE TEN COMMITMENTS

For policymakers: invest in regtech as a strategic priority, not a compliance afterthought; calibrate regulation to the development phase of the markets you govern; measure financial development outcomes rather than access metrics; design CBDC frameworks that prioritize privacy and resilience alongside efficiency; and pursue the international coordination that no single jurisdiction can achieve alone.

For financial institutions: manage the AI learning curve actively, not reactively; invest in AI governance infrastructure before deployment at scale rather than after; build the bank-fintech partnerships that preserve systemic oversight while enabling innovation; accept that the long-term commercial interest is always aligned with the customer's long-term wellbeing, because the erosion of trust is the one cost that cannot be recovered.

For fintech companies: design for engagement rather than just access; invest in financial literacy as a complement to financial product deployment; establish the transparency and data governance standards that earn rather than demand trust; and accept that the regulatory frameworks being developed are a necessary condition for the long-term sustainability of the industry, not an obstacle to it.

For everyone: understand that the financial system being built around us isn't the inevitable product of technology. It is the product of choices — choices about architecture, governance, incentives, and accountability, being made by identifiable people in identifiable institutions. Those choices can be contested, shaped, and changed. The purpose of this book is to ensure that the people making them do so with the clearest possible understanding of what is at stake.

The fintech revolution began — it is worth remembering — with a feature on a messaging app that nobody in any bank boardroom noticed. The woman in Kitui with her M-Shwari account. The Hangzhou vegetable seller with her Alipay wallet. The farmer in Ghana whose first insurance payment arrived automatically because a satellite confirmed his drought. None of these trans-formations were inevitable. Each was the product of a choice made by someone who decided that the financial system could serve more people, more effectively, than it was currently doing.

The AI inflection point presents that choice again, at a greater scale, with greater stakes. Finance, at its best, is the infrastructure of human possibility — the system that allows people to invest in the future, manage the present, and survive the past. Building that infrastructure for everyone, not just for those who were already well-served, is the most important economic challenge of our time.

The tools are there. The question is whether the will is.

Related Research — Muganyi, Yan & Sun (2021): Fintech, Green Finance, and Environmental Protection

Research published in Environmental Science and Ecotechnology (2021) documents a significant positive relationship between fintech penetration and green finance outcomes across Chinese cities: higher fintech penetration is associated with greater green bond issuance, higher green investment, and better environmental sustainability indicators.

The mechanism operates through two channels. The financial access channel: fintech makes green investment products accessible to retail investors who would never have opened a brokerage account or purchased a green bond through traditional channels. The data channel: fintech platforms generate granular behavioral data that improves the pricing of environmental risk — allowing insurance and investment products to be designed that reflect the actual distribution of climate exposure at the individual and community level.

This finding suggests that the fintech-financial development dividend extends to environmental sustainability — with direct implications for how climate finance mobilization strategies should be designed. Countries that invest in fintech infrastructure are not just improving financial access. They are building the data and distribution infrastructure that makes climate finance mobilization possible at retail scale — and in doing so, addressing the $5 trillion annual investment gap that the IEA identifies as the price of the energy transition.

Source: Muganyi, T., Yan, L. & Sun, H. (2021). Green finance, fintech, and environmental protection: evidence from China. Environmental Science and Ecotechnology 7, 100107.

CHAPTER TWELVE

Governing the AI Financial System

Central Banking at the Inflection Point

The central bank of the future won't look like the central bank of the past. It won't even look like the central bank of today. Whether it looks like something we would recognize as a central bank at all depends on the choices we make in the next decade.

147	0.3s	$23T	38%
central banks now using AI-assisted supervisory tools in some form (BIS, 2024)	average time for AI credit systems to make decisions that once took bank officers days	in assets now managed with AI input across global banking systems	of central banks report significant gaps in technical capacity to supervise AI-driven finance

Studying central banking changed how I think about regulation. Before that, I assumed the difficulty lay in the mechanics. Interest rates, credit channels, exchange rates, market expectations. These are the levers that are often discussed, the visible tools used to guide an economy. They are complex, but they are not where the real challenge sits. The difficulty runs deeper. It sits in the relationship between data, judgment, and transparency. Interacting with practitioners and regulators, I came to appreciate the level of sophistication that sits behind modern monetary policy. Advanced econometric models, vast datasets, and committees made up of highly experienced professionals all contribute to decisions that shape entire economies. These are not arbitrary choices. They are grounded in analysis, informed by history, and tested against multiple scenarios.

And yet, something about the process feels less certain than it appears.

Every model depends on the data that feeds it. Every interpretation depends on how that data is understood. The outputs may look precise, but they are built on layers of assumptions that are not always visible. There is a confidence in the system that comes from its structure, from its formality, from the rigor of the process. But beneath that, there is still judgement.

Central bank independence is often treated as a given. It is presented as a safeguard, a way to ensure that decisions are made without undue influence. In practice, independence is more nuanced. These decisions are made by people. They bring their experience, their instincts, and their interpretation of the data into the room. They rely on what they have seen work before. They rely on patterns that have held up over time. That does not make the system weak. It makes it human. At the same time, it introduces a layer of opacity that is difficult to fully address. Not because information is hidden, but because interpretation cannot be separated from the individuals making the decision. Two equally qualified experts can look at the same dataset and arrive at different conclusions. Both can justify their reasoning. Both can be grounded in sound analysis.

That is the nature of complex systems.

The Bank of England, also less commonly known as the old lady of Threadneedle Street, a nickname from a 1797 satirical cartoon by James Gillray was founded in 1694 to finance a war. It lent £1.2 million to the Crown at eight percent interest, received a royal charter in return, and began the long accumulation of institutional authority that would eventually make it the model for central banks across the world. For over three centuries, its primary responsibilities—regulating the money supply, determining interest rates, and acting as lender of last resort during periods of financial instability—have remained clearly defined. Difficult, important, and requiring exceptional judgement, certainly. But legible. The instruments were understood. The transmission mechanisms, if not perfectly predictable, were at least theoretically tractable.

The AI inflection point is doing something to central banking that no previous financial innovation has managed: it is making the transmission mechanisms genuinely opaque. Not opaque in the familiar sense of complex and hard to model — central bankers have always operated in conditions of radical uncertainty. Opaque in the deeper sense that the systems through which monetary policy now travels into the real economy are partly composed of algorithms whose responses to rate changes, liquidity conditions, and forward guidance cannot be predicted from first principles, because those algorithms were not designed to respond to first principles. They were trained to respond to patterns. And the patterns of the AI era are unlike anything the training data of the pre-AI era contained.

This is the central banking problem of our time: not the familiar challenge of calibrating the right interest rate under uncertainty, but the new challenge of governing a financial system whose behaviour can no longer be fully traced, whose decision logic cannot be fully interrogated, and whose failure modes cannot be fully anticipated — because those failure modes emerge from the interaction of systems that no single institution designed and no single regulator oversees.

THE MONETARY POLICY TRANSMISSION PROBLEM

Monetary policy transmission — the mechanism by which a central bank's decisions about interest rates and money supply propagate through the financial system and eventually affect employment, output, and prices in the real economy — has always been imprecise. The channels are well-mapped in academic literature: the bank lending channel, in which rate changes affect the cost and availability of credit; the asset price channel, in which rate changes affect equity and property valuations; the exchange rate channel, in which rate changes affect the cost of imports and the competitiveness of exports; and the expectations channel, in which forward guidance shapes the beliefs of businesses and households about the future, affecting their decisions today.

Each of these channels now passes through AI. Lending decisions that once reflected a bank officer's judgement about a borrower's creditworthiness now emerge from models trained on historical data. Asset valuations that once reflected the considered views of fund managers now incorporate AI-generated signals that detect patterns invisible to human analysts. Exchange rate movements that once tracked the slow-moving fundamentals of trade balances and inflation differentials now incorporate microsecond algorithmic responses to central bank communications. And expectations formation — perhaps the most important channel in the modern monetary framework — now occurs in an information environment saturated with AI-curated content, AI-generated analysis, and AI-assisted trading that can amplify or suppress the signals that central banks intend to send.

The consequence is that the relationship between a central bank's policy decision and its eventual effect on the economy has simultaneously become both faster and less predictable. Faster because AI systems respond to policy signals in milliseconds where human actors responded in days or weeks. Less predictable because the aggregate behavior of multiple interacting AI systems — each trained on different data, optimizing for different objectives, operating at different time horizons — cannot be reliably forecast by the models that central banks use to project the effects of their decisions.

The Bank for International Settlements has described this phenomenon as the 'AI amplification problem': the tendency of AI-mediated financial markets to amplify both the speed and the volatility of monetary policy transmission, creating a world in which rate changes hit their intended targets faster than the models predicted but also create unintended consequences that the models did not anticipate. The 2022 UK gilt market crisis — in which liability-driven investment strategies, many of them algorithmically managed, created a self-reinforcing spiral of selling that required Bank of England intervention within days of the Truss government's mini-budget — was an early and costly preview of what AI amplification looks like in practice.

THE DATA REVOLUTION AND ITS DISCONTENTS

The same technology that is complicating monetary policy transmission is simultaneously offering central banks tools for economic intelligence that would have been unimaginable a generation ago. Real-time data — from payment systems, satellite imagery, mobile phone networks, and social media — creates the possibility of monitoring economic activity at a granularity and speed that makes the traditional statistical infrastructure of central banking look like archaeology by comparison.

The nowcasting revolution — using high-frequency data to estimate the current state of an economy before official statistics are compiled and published — has been one of the genuine success stories of AI in central banking. The Federal Reserve's Atlanta Fed GDPNow model, which uses over forty data series to produce a real-time estimate of US GDP growth, has consistently outperformed the professional consensus in the weeks before official data releases. The Bank of England's equivalent, developed in partnership with the Alan Turing Institute, can detect economic turning points in real-time payment data weeks before they appear in conventional survey measures. During the COVID-19 pandemic, these nowcasting tools provided policymakers with economic intelligence that traditional data releases — which lag by months — could not have delivered.

But the data revolution creates new problems alongside the ones it solves. The first is the problem of data governance: whose data is it, who controls access to it, and on what terms? The real-time payment data that provides the most economically informative signal of consumer behavior is held by private payment processors — Visa, Mastercard, Stripe, and their equivalents — who provide access on commercial terms that can be changed, restricted, or revoked. Central banks that have built their real-time intelligence infrastructure on privately held data are creating dependencies on commercial counterparties that raise governance questions that no central bank statute was designed to address.

The second problem is the difference between data abundance and understanding. Having more data, faster, doesn't automatically produce better economic understanding. The patterns that AI systems identify in high-frequency data are often statistically robust but economically uninterpretable: the model can tell you what is happening but not why, which limits its value for the kind of structural analysis that sound monetary policy requires. A central bank that raises interest rates because its AI system detected a concerning pattern in card payment data, without understanding the economic mechanism that pattern represents, is making policy based on correlation rather than causation — a form of epistemic risk that the pre-AI statistical framework, for all its limitations, was at least designed to guard against.

SUPERVISORY INTELLIGENCE: THE CAPACITY GAP

The governance challenge of AI in finance is not primarily a rule-writing problem. The rules — from the EU AI Act to the FSB's principles for AI governance — are being written. The more urgent problem is the capacity gap between the analytical capability of the supervised institutions and the analytical capability of the supervising institutions.

Consider what a sophisticated AI-enabled bank looks like from the outside, as a supervisory subject. Its credit models are proprietary systems with hundreds of millions of parameters. Its trading algorithms operate at microsecond speeds. Its fraud detection systems process millions of transactions daily using techniques that their own developers cannot fully explain. Its risk management infrastructure integrates data from dozens of sources via continuously updated machine learning pipelines. The institution itself employs hundreds of data scientists, machine learning engineers, and AI researchers to build and maintain these systems.

Now consider the supervisory capacity that a typical financial regulator can bring to bear on this institution. Annual examination cycles. Sampled transaction reviews. Document-based assessments of model governance. Risk-based supervisory programs designed for a financial system whose riskiest decisions were made by humans in meetings and recorded in paper files. The mismatch is not a matter of regulatory will or even regulatory resources at the margin. It is a structural gap that has opened between the technology of supervised institutions and the technology of supervising institutions — a gap that, if it continues to widen, will eventually make meaningful supervision of the AI-driven financial system impossible.

The most sophisticated regulatory responses to this gap are investing in supervisory technology at a scale and seriousness that matches the investment being made in the supervised systems themselves. The FCA's data science unit — which in 2024 employed over 150 data scientists, more than many of the fintech firms it supervises — represents one model. The ECB's supervisory technology program, which uses machine learning to monitor the 112 significant institutions under its direct oversight for early signs of financial stress, represents another. The Monetary Authority of Singapore's Project MindForge, which brings together a consortium of banks, insurers, and capital market firms to develop a shared AI risk management framework for the financial sector, represents a third.

What these program share is a recognition that supervising AI with non-AI tools is not a sustainable strategy. The economics of supervisory capacity are, in this respect, inexorable: the ratio of supervised complexity to supervisory capacity will continue to widen unless regulators invest in the same class of technology they are seeking to govern.

The Bank of England's AI-Driven Supervisory Intelligence Program

Since 2019, the Bank of England has progressively integrated machine learning and AI into its supervisory and financial stability operations. Joint surveys conducted with the Financial Conduct Authority in 2019, 2022, and 2024 have tracked the rapid adoption of AI across the UK financial services sector, with around three-quarters of surveyed firms reporting some form of AI or machine-learning use by 2024. The Bank's own internal deployment of machine learning tools spans systemic risk monitoring, text analysis of supervisory correspondence, and labor market modelling — using techniques including natural language processing, network analytics, and anomaly detection across regulatory filings, market data, and supervisory intelligence.

The September 2022 gilt market crisis — triggered by the liability-driven investment (LDI) unwind following the UK government's fiscal announcement — underscored the value of these capabilities. The Bank's data infrastructure and monitoring tools enabled staff to identify stress in gilt repo markets and prepare the emergency gilt purchase intervention that stabilized the market. The Financial Policy Committee subsequently identified AI-powered supervisory surveillance as a strategic priority, noting that the effective monitoring of AI-related risks to financial stability would require flexible, forward-looking approaches blending quantitative and qualitative intelligence.

The Bank has been explicit that its AI tools are human augmentation instruments, not autonomous decision-making systems. In 2025, it launched an AI Consortium of private-sector firms and AI experts to deepen understanding of both AI's benefits and its risks, and to assess whether existing regulatory frameworks adequately enable safe AI adoption. The Bank has also been candid about the limitations of its approach: machine learning models are sensitive to the patterns in their training data and may miss novel stress events that look unlike anything in the historical record. The 2022 LDI crisis had historical precedents that the data contained. The next crisis may not.

CENTRAL BANK DIGITAL CURRENCIES AND THE AI GOVERNANCE NEXUS

AI and central bank digital currencies represent one of the most important and least adequately governed frontiers in the future of money. CBDCs, as discussed in Chapter Five, are digital forms of state money — direct liabilities of central banks, accessible to the public through digital infrastructure. AI enters this picture at multiple points: in the design of CBDC systems, in their operation, in their surveillance implications, and in the governance challenges they create for the institutions that issue them.

The design dimension is the most immediate. CBDC systems that use AI to process transactions, detect fraud, manage monetary policy transmission, and calibrate distribution to different economic sectors are being built in over 100 countries simultaneously. The People's Bank of China's digital yuan — the e-CNY — already incorporates AI-driven transaction monitoring that can identify suspicious patterns in real time and flag them for regulatory review without human initiation. The European Central Bank's Digital Euro proposals include AI-powered compliance checking that would automate anti-money-laundering assessments for transactions above defined thresholds.

These applications are efficient. From a governance perspective, they are also unprecedented. When a CBDC transaction is flagged, blocked, or reversed by an AI system on behalf of a central bank that is ultimately an arm of the state, the rights implications are categorically different from those that arise when a private commercial bank makes the same decision. The central bank is more than a financial intermediary; it is the state's monetary authority. Its decisions about who can transact, on what terms, and subject to what monitoring are decisions about the exercise of sovereign monetary power — and the governance models appropriate for that power are different from those appropriate for private fintech systems.

The BIS's Project Aurora — which is exploring how AI can be used in CBDC systems to detect and prevent money laundering without compromising transaction privacy — represents the most technically ambitious attempt to navigate this tension. The project uses privacy-preserving computation techniques to allow AI fraud detection systems to operate on encrypted transaction data, identifying suspicious patterns without the monitoring institution being able to read the content of individual transactions. If successful, this approach could achieve the compliance objectives of CBDC surveillance without the privacy cost — a design solution that the political legitimacy of CBDC adoption may require.

THE MODEL RISK BLIND SPOT

The single most underappreciated governance risk in the AI-driven financial system is what the author's own research identifies as the early-phase risk elevation described in the inverted U framework of Chapter Nine. At the systemic level, this dynamic creates a period during which the aggregate riskiness of the banking system is rising — driven by competitive AI adoption pressures — while the supervisory frameworks meant to detect this rise are still calibrated to a pre-AI risk environment.

The specific manifestation of this risk that has attracted the most serious concern from financial stability authorities is model herding: the tendency of financial institutions adopting similar AI architectures to produce correlated behavior that amplifies rather than diversifies systemic risk. When every major bank's credit model is a gradient-boosted tree trained on similar macroeconomic variables, the models will agree in their assessments of credit conditions. When conditions are good, they will collectively expand credit. When conditions deteriorate, they will collectively contract. The cyclical amplification this produces — more credit in booms, less in busts — is precisely the pattern that macroprudential regulation was designed to counteract, now being recreated by the technology that was supposed to improve credit assessment.

The IMF's Global Financial Stability Report of 2023 documents this concern with unusual directness: AI-driven correlation in credit decisions could increase the amplitude of credit cycles by 15 to 20% relative to a counterfactual financial system without AI model herding. At the scale of the global credit system, a 15% amplification of credit cycles represents a very large increase in the probability of the kind of credit-boom-and-bust that produced the 2008 financial crisis.

The governance response to model herding is conceptually clear but practically difficult: require diversity in AI model architectures across institutions, mandate stress testing that specifically probes for shared model vulnerabilities and build the supervisory capability to assess AI model similarity across the financial system. Implementing these requirements demands technical expertise in AI model assessment that most financial regulators are still building — and doing so against a timeline set by the pace of AI adoption in the institutions they supervise, rather than by the pace of their own capacity development.

Research Framework: The Three Governance Gaps

Analysis of the intersection between the author's Eurozone bank risk research and the BIS's AI governance framework identifies three structural governance gaps that no existing regulatory framework has fully closed.

The first is the speed gap: regulatory cycles operate over months and years; AI systems make serious decisions in milliseconds. The governance tools designed for the former — annual examinations, periodic model validations, quarterly reporting — cannot catch problems that propagate at the speed of the latter. Closing this gap requires real-time supervisory technology, not faster application of existing tools.

Then there is the interpretability gap: the most accurate AI models used in finance are also the least interpretable. Regulatory requirements for explainability — the ability to provide comprehensible accounts of why an AI system made a specific decision — create a systematic incentive to use less accurate but more explainable models, or to provide post-hoc rationalizations of model outputs that don't accurately describe the model's actual decision process. Neither outcome serves the governance purpose.

The third is the jurisdictional gap: AI systems deployed in finance are global; the regulatory frameworks governing them are national. A credit model trained in the United States can be deployed to serve borrowers in fifty countries from a single server farm, subject to the regulatory requirements of the jurisdiction where it was trained and potentially none of the requirements of the jurisdictions where its consequences are felt. Closing this gap requires international regulatory coordination of a kind and speed that the history of financial regulation suggests is difficult to achieve before, rather than after, a crisis demonstrates its necessity.

Sources: Muganyi et al. (2022), Financial Innovation; BIS Working Papers 1194 (2024); IMF Global Financial Stability Report Chapter 3 (2023); FSB Artificial Intelligence and Machine Learning in Financial Services (2023)

CHAPTER THIRTEEN

The Dollar Under Siege

BRICS, Geopolitics, and the Future of Reserve Currency

The dollar's exorbitant privilege was never permanent. It was the product of specific historical circumstances — the destruction of European and Japanese industrial capacity in 1945, the Bretton Woods settlement, the oil shock of 1973 — that are now four to eight decades in the past. History does not repeat itself, but monetary history has a particular tendency to rhyme.

58%	$7-9T	50+	40%
of global foreign exchange reserves held in US dollars, down from 71% in 2000 (IMF COFER, 2024)	*in daily global foreign exchange market volume, 88% involving the dollar in one leg*	*countries now conducting some bilateral trade outside the dollar system*	*of global GDP represented by BRICS+ members as of 2024 expansion*

In August 2023, in Johannesburg, the leaders of Brazil, Russia, India, China, and South Africa announced the most significant expansion of BRICS since the grouping was formalized in 2009. Six new members — Saudi Arabia, Iran, Ethiopia, Egypt, Argentina, and the United Arab Emirates — were invited to join. The announcement was accompanied by the language of historic transformation: a new multipolar world order, a challenge to the dollar-dominated financial system, a BRICS currency that would free the global south from the tyranny of monetary dependence on a single nation's policy decisions.

The reality, as it nearly always is in international monetary affairs, was considerably more complicated than the rhetoric. No BRICS currency emerged from Johannesburg, and the technical obstacles to creating one — which are substantial — were not resolved by the political momentum of the summit. Argentina subsequently elected a government that explicitly abandoned BRICS membership aspirations in favor of closer dollar alignment. The geopolitical coherence of a grouping that includes both India and China — two nuclear powers with unresolved border disputes and competing regional ambitions — remained as questionable as ever.

And yet something real is happening in the global monetary system. The dollar's share of global foreign exchange reserves has fallen from 72% in 2001 to 57% in 2026 — a 15-percentage point decline that represents the largest relative shift in reserve currency composition since the collapse of Bretton Woods. Bilateral trade between China and Russia is now settled almost entirely in renminbi and rubles rather than dollars. Saudi Arabia has accepted renminbi payment for a portion of its oil sales to China, breaking a 50-year convention that crude oil is priced and settled in dollars. And the sanctions imposed on Russia following the 2022 invasion of Ukraine have demonstrated, to a watching world, that dollar-denominated financial infrastructure can be weaponized against sovereign states in ways that create powerful incentives for those states — and for others observing their treatment — to develop alternatives.

THE ARCHITECTURE OF DOLLAR DOMINANCE

Understanding the pressure the dollar faces requires that we first understand how dollar dominance was constructed and what maintains it. The dollar's reserve currency status rests on four reinforcing foundations that have, in combination, made it functionally irreplaceable for global commerce for eighty years.

The first is network effects. International trade is invoiced in dollars, not because the United States requires it, but because the convention is self-reinforcing: if most of your trading partners accept and prefer dollars, the transaction cost of using any other currency rises. Every bilateral trade relationship in which a non-dollar currency is used requires currency conversion, hedging, and the maintenance of foreign exchange reserves in that currency — costs that disappear when both parties use the same reserve currency. Overthrowing a currency network effect requires not just a better alternative but a coordinated shift by enough participants simultaneously to make the alternative viable — a coordination problem of extraordinary difficulty.

The second is financial market depth. The US Treasury market — with over $30 trillion in outstanding securities and the deepest, most liquid secondary market in the world — provides the safe asset that global reserve managers require. No other sovereign bond market comes close in depth, liquidity, or the reliability of access under stress. In every episode of global financial stress since 1945, capital has flowed into U.S. dollar assets as a safe haven — a pattern so consistent that it has become self-fulfilling: the dollar is safe because everyone believes it is safe; everyone believes it is safe because every crisis has confirmed the belief.

The third is the dollar's role in commodity pricing. Crude oil, natural gas, copper, gold, wheat — the commodities that underlie the physical economy of global trade — are priced and settled in dollars. This means that every nation that imports energy or food must maintain dollar reserves to fund those imports, regardless of whether its own economy or trading relationships are primarily dollar-denominated. The petrodollar system, formalized through US-Saudi arrangements in the 1970s, created a structural source of dollar demand that has persisted for half a century and that its critics argue amounts to a hidden subsidy to US monetary policy — the ability to run current account deficits that would be unsustainable for any other currency.

A fourth dimension: SWIFT and the dollar-denominated financial infrastructure that underlies global payments. The Society for Worldwide Interbank Financial Telecommunication processes the messaging for the vast majority of international bank transfers, and the clearing of dollar transactions passes through the US financial system, giving the United States visibility into, and leverage over, the financial flows of every country that participates in the global economy.

SANCTIONS AS ACCELERANT: THE RUSSIA CASE

The decision to exclude Russia from SWIFT and freeze approximately $300 billion of Russian central bank assets held in Western jurisdictions following the 2022 Ukraine invasion was the most serious exercise of dollar weaponization in the history of the international monetary system. It was also, in an important sense, a demonstration that shook not just Russia but every country that observes its own foreign exchange reserves as a potential liability rather than an asset if political circumstances change.

The freezing of Russian central bank assets — a legally and strategically novel step that went beyond anything previously attempted against a major economy — demonstrated that dollar reserves held in Western institutions are not unconditionally available to their nominal owners. They are available at the discretion of the Western governments that control the financial infrastructure in which those reserves sit. For countries with relationships with the West that are cooperative but not alliance-based — the Gulf states, major Asian economies, much of the global south — the Russia episode raised a question that had not previously been live: what assurance do we have that our reserves will not be treated the same way if our relationship with Washington deteriorated?

The answer, for many of these countries is: not enough. And the search for alternatives — conducted with far less ideological heat than the BRICS summit rhetoric suggested, but with considerably more practical urgency than the Western financial press acknowledged — has accelerated.

Russia's own response to the SWIFT exclusion illustrates both the limitations of the dollar alternative that exists today and the direction of travel. Russia's SPFS (System for Transfer of Financial Messages, its domestic SWIFT alternative) had 469 participants at the time of the sanctions and processed approximately 20% of domestic Russian financial message traffic. It has since been expanded to cover bilateral payments with China, India, and several Central Asian economies. China's CIPS (Cross-Border Interbank Payment System), designed as an international alternative to SWIFT for renminbi-denominated transactions, has seen transaction volumes triple since 2022 and now covers over 100 countries. Neither system replaces SWIFT. Both demonstrate that the infra-structure for a partial alternative is being built.

THE RENMINBI CHALLENGE: REAL AND OVERSTATED

The Chinese renminbi is the most plausible long-run challenger to the dollar's reserve currency status, and the challenge it poses is simultaneously more real and more overstated than most commentary acknowledges.

Renminbi internationalization is becoming tangible as structural conditions are being put in place. In 2014, China's GDP exceeded that of the US by purchasing power parity and is still catching up on market exchange rates. China's trade volumes — the largest in the world — create a structural rationale for renminbi invoicing that did not exist when China was a developing export economy. The e-CNY — China's central bank digital currency — is being explicitly designed with cross-border functionality, and mBridge, the BIS Innovation Hub project connecting the digital currency systems of China, Hong Kong, the UAE, and Thailand, has demonstrated that cross-border CBDC settlement can operate without the dollar-denominated correspondent banking system.

More overstated because the most fundamental precondition for a reserve currency — full capital account convertibility, which allows holders of renminbi to freely move their money into and out of China and into any other currency without restriction — remains absent, and its absence is not a technical oversight but a deliberate policy choice. Capital controls are the mechanism through which the People's Bank of China manages exchange rate stability and insulates the domestic financial system from external shocks. Removing them would expose the Chinese financial system to volatility in capital flows that its current institutional structure is not designed to manage. Every time China's economy has shown signs of stress — 2015, 2019, 2022 — the capital controls have been tightened, not loosened.

The renminbi's share of global foreign exchange reserves was 2.7% in 2024 — up from essentially zero in 2010, but still a fraction of the dollar's 58% and below the euro's 20%. The arc is upward. The pace is slow. And the political conditions in China — which would require a fundamental redesign of the relationship between the state and the financial system to enable full convertibility — don't currently favor the acceleration that reserve currency displacement would require.

Project mBridge is one of the most technically advanced multi-CBDC platforms in existence, and one of the least discussed in Western financial media. Developed under the auspices of the BIS Innovation Hub in partnership with the central banks of China, Hong Kong, the UAE, and Thailand, mBridge allows participating institutions to settle cross-border transactions directly in their own CBDCs, without routing through the dollar-denominated correspondent banking system that underlies conventional international payments.

In its 2022 pilot, mBridge settled $22 million in cross-border transactions among 20 commercial banks across four jurisdictions in a matter of seconds, at a fraction of the cost of equivalent conventional transfers. The transactions included real commercial flows — including trade-related finance, foreign direct investment settlements, and energy payments — demonstrating that the system can handle genuine commercial complexity, not just proof-of-concept toy transactions.

The geopolitical significance of mBridge extends beyond its current scale. If the platform is extended to cover Saudi Arabia's oil sales to China — a scenario being actively explored, following Saudi Arabia's invitation to join BRICS — it would create a mechanism for settling a portion of the global oil trade entirely outside the dollar system. The volumes involved would not, initially, be large enough to materially affect dollar reserve requirements. But the precedent — of oil trade settled in CBDCs without dollar intermediation — would be historically significant in a way that the transaction volumes alone don't capture.

The United States has noted mBridge's development with concern. The Treasury Department's 2023 report on digital assets identified multi-CBDC platforms as a potential mechanism for sanctions evasion and dollar system circumvention. Whether this concern translates into policy action — and what that action might look like — is one of the most consequential unresolved questions in the geopolitics of money.

THE BRICS CURRENCY QUESTION: WHY IT IS HARDER THAN IT SOUNDS

The proposal for a BRICS common currency — variously described as the R5 (for the initial currencies of member states: real, ruble, rupee, renminbi, rand), the Unit, or simply the BRICS currency — is one of those ideas whose political appeal far exceeds its practical tractability. Understanding why requires a brief engagement with the economics of currency unions, which has a somewhat dispiriting academic record.

A currency union among countries with different inflation rates, different business cycles, different levels of financial development, and different political systems requires either a mechanism for managing the divergences that will inevitably open up among members or the willingness of some members to subordinate their economic policy to the needs of the union. The euro demonstrates both the possibility and the cost of a currency union among countries with differing economic structures: the divergence between German and southern European economic performance in the decade after the 2008 crisis was, in significant measure, a consequence of sharing a currency without sharing the fiscal and political union that would allow the divergences to be managed.

BRICS presents this challenge in a more extreme form. The member states have wildly divergent inflation rates (Brazil's has averaged 7% over the past decade; China's has averaged 2%), different political systems that create different tolerance for the economic discipline that a currency union requires, and strategic interests that are frequently in conflict. India and China have gone to war and continue to have unresolved border disputes. Brazil's foreign policy alternates between engagement and withdrawal from multilateral commitments with each election cycle. Sanctions have thoroughly restructured Russia's economy. These are not the conditions in which a common monetary policy is feasible.

The more realistic BRICS monetary project is not a common currency but a common settlement mechanism — a platform that allows BRICS members to conduct trade among themselves without routing through the dollar system, using their own national currencies for bilateral pairs and a notional unit for multilateral clearing. This more closely reflects what the 2023 Johannesburg summit actually discussed, and aligns more directly with what mBridge is building. It does not threaten dollar dominance in the short run. Over a decade or more, if adopted at the scale of intra-BRICS trade flows, it could reduce the dollar demand generated by that trade by a meaningful amount.

FINTECH, AI, AND THE GEOPOLITICS OF MONEY

The intersection of the dollar-dominance question with the AI and fintech revolution documented in these chapters creates dynamics that have received insufficient attention in either the financial economics or the international relations literature.

The most consequential of these dynamics is the role of AI-powered fintech in enabling the dollar alternative infrastructure that has, until recently, been technically impossible to build at scale. Cross-border payments between countries with under-developed correspondent banking relationships have historically been slow, expensive, and dependent on dollar intermediation exactly because the infrastructure for direct bilateral settlement did not exist. AI-powered payment rails — which can manage liquidity, assess counterparty risk, and execute FX conversion in real time across currency pairs with limited market depth — are creating the technical possibility of bypassing that dependence in ways that would not have been achievable with pre-AI technology.

The digital yuan's cross-border capability is the most advanced example. The e-CNY system uses AI-driven real-time risk management to manage liquidity and settlement risks in cross-border transactions in a currency with capital controls — a technical feat that the conventional wisdom in the international monetary economics had held to be impossible. You cannot, long-standing theory suggests, internationalize a currency with capital controls, because the controls prevent the offshore holdings of that currency from being freely used. The e-CNY's programmable money features — which allow transactions to be executed within defined parameters set by the issuing authority — offer a way around this constraint: capital controls can be embedded in the currency itself, limiting the uses to which offshore renminbi can be put rather than limiting its circulation.

Whether this technical innovation is sufficient to overcome the fundamental economic constraints on renminbi internationalization is a question that events over the next decade will answer. What is clear is that the AI and fintech revolution has changed the technical frontier of what is possible in international monetary architecture, and that the geopolitical actors who are most motivated to change the existing architecture are investing heavily in exploring those new possibilities.

> ### The De-Dollarization Spectrum: What the Data Actually Shows
>
> The de-dollarization debate is frequently conducted at a level of abstraction that obscures what the data actually shows. A more precise account distinguishes between four different dimensions of dollar centrality, each of which is moving at a different pace and in a different direction.
>
> Reserve composition: The dollar's share of global reserves has fallen from about 72% in 2001 to 57% in 2026. The pace of decline has accelerated since 2022. The beneficiaries have been non-traditional reserve currencies — the Australian and Canadian dollar, the Swedish krona, the Korean won — rather than the renminbi, which has captured only about 2.7% of the shift.
>
> Trade invoicing: The dollar's share of global trade invoicing has been more stable than its reserve share, remaining above 50% even as its reserve share fell. However, bilateral trade between China and its major partners is increasingly invoiced in renminbi, and the dollar's share of commodity invoicing is under more sustained pressure than its share in manufactured goods trade.
>
> Payment messaging: SWIFT dollar transaction volumes have remained broadly stable in absolute terms even as CIPS and bilateral alternatives have grown, suggesting that the alternatives are currently adding capacity for non-dollar flows rather than displacing existing dollar flows. This could change if sanction risks cause more countries to route their dollar transactions through alternative infrastructure.
>
> Financial market depth: No challenger has emerged to seriously contest the US Treasury market's role as the global safe asset. This dimension of dollar centrality is the most structurally entrenched and the most resistant to the geopolitical and technological pressures affecting the others.
>
> *Sources: IMF COFER Database Q4 2024; BIS Triennial Central Bank Survey 2022; SWIFT RMB Tracker 2024; Federal Reserve Board international finance discussion papers*

The purpose of this chapter has been to examine the claim that the U.S. dollar is "under siege." Much of the commentary surrounding the future of the dollar is loud, confident, and often misleading. When the evidence is examined carefully, the picture is more nuanced.

What we are likely to see in the coming decades is not the collapse of dollar dominance, but the gradual emergence of a more fragmented global financial system. Certain international transactions will increasingly be settled in alternative currencies. The Chinese renminbi will expand its role in trade settlement, where China is the primary economic partner. Regional initiatives — including the possibility of a BRICS-aligned payment framework — may modestly expand the use of non-dollar currencies in cross-border trade.

Yet these developments should not be mistaken for a fundamental challenge to the dollar's global reserve status. The structural foundations that support the dollar — deep and liquid capital markets, the scale of US financial institutions, legal credibility, and the central role of US Treasury securities in the global financial system — remain unmatched. No existing currency currently replicates this institutional ecosystem.

What is changing is not the center of gravity of the global financial system, but its edges. The future will likely be one in which a greater share of regional trade is settled outside the dollar, while the core architecture of global finance continues to run through it.

The dollar, in other words, is not disappearing. It is operating within a system that is slowly becoming more plural — but not yet truly multipolar.

CHAPTER FOURTEEN

Innovation or Stagnation

The EU AI Act, Fintech Investment Flows,
and the Digital Wallet Wars

Europe builds the rules. America builds the companies. China builds the infrastructure. The question for the next decade is whether any of these strategies produce both safety and prosperity — or whether the pursuit of one systematically undermines the other.

5.2B+	14%	€17B
digital wallet users globally projected by 2028, up from 3.4B in 2024 (Juniper Research)	EU share of global fintech investment in 2023; down from 27% by 2019 (KPMG Pulse)	estimated EU fintech compliance cost burden from AI Act and DORA combined by 2027

When the EU AI Act was passed, I remember getting a LinkedIn notification that seemed to capture the mood of the moment. The post that had gone viral showed a photo of European regulators celebrating after the parliamentary vote. The caption read: "They regulate everything and innovate nothing." It was the kind of line that travels quickly. It was sharp, provocative, and easy to repeat. It was also incomplete. Still, it struck a nerve. Some founders I know shared it. Some agreed with it outright. Others pushed back, but even then, the frustration was visible. Regulation, in their eyes, had become something that slowed things down, something that stood in the way of building, testing, and scaling new ideas.

I understood where that reaction came from.

Innovation moves quickly. It thrives on iteration, on the ability to try something, learn from it, and move again without too much friction. When that process is interrupted, it can feel like momentum is being lost. For someone building a company, that loss of momentum is not abstract. It affects decisions, timelines, and in some cases, survival. At the same time, I could not ignore another reality.

Regulation has shaped many of the systems we now take for granted. It is easy to overlook this because it often works quietly, in the background, influencing design in ways that does not draw attention to themselves. The shift to USB-C charging across devices is one example. It did not emerge purely from market forces. It was guided, in part, by regulatory pressure aimed at standardization and consumer convenience. That outcome feels obvious now, but it was not inevitable.

The relationship between innovation and regulation is not fixed. It changes depending on how both sides evolve. Regulation can slow things down, but it can also set direction. It can create constraints, but those constraints can shape how solutions are built. In some cases, they can even create the conditions for new forms of innovation to emerge.

What complicates this further is the speed at which technology is developing. Artificial intelligence is moving at a pace that challenges traditional approaches to oversight. New models, new applications, and new use cases appear faster than most regulatory processes can respond. This creates tension. On one side, there is a need to manage risk, protect users, and maintain stability. On the other, there is a need to allow space for experimentation and growth. That tension is not new, but it feels more pronounced now.

The EU was the first major jurisdiction in the world to impose wide-ranging legal requirements on the development and deployment of artificial intelligence systems. The achievement was, by any measure, historic: the first attempt by any legislative body to govern AI not as an afterthought or a sector-specific patch but as a comprehensive framework addressing risk, rights, transparency, and accountability across the full range of AI applications.

Within forty-eight hours, the commentary from the European fintech industry was scathing. Trade associations representing more than two thousand financial technology companies across the EU issued statements warning that the Act's requirements for high-risk AI systems would impose compliance burdens that would make European fintech companies uncompetitive relative to their US and Asian counterparts, drive investment out of European markets, and push the most innovative AI development offshore. The argument was not new — it had been made throughout the Act's five-year legislative journey — but its forcefulness at the moment of enactment reflected real anxiety rather than mere lobbying posture.

The question this chapter addresses is whether that anxiety is justified. Not meaning that industry participants want the answer adjudicated — they have a direct interest in their preferred conclusion — but in the sense that the empirical evidence about regulatory cost, innovation, and competitive dynamics in global fintech markets allows an honest assessment. The answer is more nuanced and more troubling for the EU than either the Act's enthusiasts or its critics have acknowledged.

WHAT THE EU AI ACT ACTUALLY REQUIRES OF FINTECH

The EU AI Act's risk classification system divides AI applications into four categories: unacceptable risk (prohibited), high risk (heavily regulated), limited risk (transparency requirements only), and minimal risk (essentially unregulated). For financial services, the classification that matters most is high risk, which encompasses credit scoring, insurance pricing, claims assessment, and certain fraud-detection applications.

For each high-risk AI application, the Act imposes a suite of requirements that are individually defensible but collectively demanding. Technical documentation must be maintained describing the system's architecture, training data, performance metrics, and known limitations. A conformity assessment must be conducted before deployment — either internally by the deploying institution or, for certain applications, by a notified body (an accredited independent assessor). The system must be registered in an EU-wide AI database. Human oversight mechanisms must be implemented, including the ability for trained humans to override AI decisions in high-stakes cases. The system must be monitored continuously in deployment, with post-market surveillance requirements. Affected individuals must be informed when AI is used to make consequential decisions about them. And they have the right to receive an explanation of those decisions in terms they can meaningfully engage with.

The compliance cost of these requirements is not trivial. An independent assessment commissioned by the European Parliament estimated that the conformity assessment process alone would cost between €6,000 and €7,500 per application for AI systems subject to third-party assessment — a figure that scales with the number of distinct AI applications a fintech deploys. For a mid-sized European neobank with fifteen to twenty

AI-enabled products and features, the one-time conformity cost runs from €100,000 to €150,000 before ongoing monitoring and documentation obligations are included.

For the largest European financial institutions, these costs are material but manageable — they represent a fraction of the technology and compliance budgets that major banks already maintain. For the European fintech sector's demographic backbone — the seed-stage and Series A companies that do not yet have large compliance functions, which are deploying AI as a core competitive tool rather than a supplement to existing operations, and which are operating on cash runways measured in months rather than years — the compliance architecture of the AI Act represents a structural disadvantage relative to competitors in jurisdictions with less demanding requirements.

THE BRUSSELS EFFECT IN PRACTICE: WHAT HISTORY SUGGESTS

The 'Brussels Effect' — the tendency of EU regulation to become the de facto global standard because the cost of maintaining different product versions for different jurisdictions exceeds the cost of complying with the strictest available standard everywhere — is the optimist's case for the AI Act's global consequences. If the EU's AI governance framework becomes the template that global companies adopt universally to avoid the cost of jurisdiction-specific compliance, the Act's requirements for transparency, explainability, and human oversight will propagate through the global AI industry regardless of what other regulators do.

The Brussels Effect worked, powerfully, for data protection. The General Data Protection Regulation of 2018 became the data governance standard globally not because other regulators adopted it but because the companies that needed to serve EU customers adopted it globally — and once they had done so, the marginal cost of applying the same standards elsewhere was low. GDPR's requirements are now embedded in the privacy policies, data architectures, and compliance program of companies headquartered in California, Singapore, and Tokyo, none of which are required by their home regulators to follow them.

The issue at hand is whether AI regulation can be compared to data regulation such that similar regulatory mechanisms could be applied. The available evidence presents a nuanced perspective. On the positive side: the largest AI companies — OpenAI, Google DeepMind, Meta AI, Anthropic — are investing substantially in compliance with the AI Act's requirements for their EU operations, and several have indicated that their responsible AI frameworks will be applied globally rather than only in the EU. On the negative side, the AI Act's explainability requirements impose technical constraints — not just process and documentation constraints — on models that can be used for high-risk applications. A company that cannot deploy its most accurate model in the EU because that model fails the explainability test cannot solve this problem through global policy harmonization. It faces a genuine capability constraint that does not apply to its competitors in less-regulated markets.

This is the AI Act's most significant competitive consequence, one that the Brussels Effect optimism does not adequately address. The GDPR imposed costs and required changes in process; it did not prevent companies from building the most capable data products. The AI Act's high-risk AI provisions for finance may prevent European financial institutions from deploying the most capable AI models — not because of cost or process, but because those models cannot meet the explainability requirement mandated by European law.

WHERE THE INVESTMENT IS GOING

Global fintech investment peaked at $238 billion in 2021, collapsed to $91 billion in 2023 as rising interest rates ended the era of cheap capital that had sustained loss-making hypergrowth strategies, and began a selective recovery in 2025. The geography of that recovery is as revealing as its aggregate scale.

The United States remains the dominant destination for global fintech investment, absorbing approximately 45% of global capital in 2023 despite representing only 4% of the world's population. The concentration reflects the depth of US capital markets, the density of fintech-specialist investors, the regulatory environment that enables rapid product deployment and iteration, and the size of the domestic market, which makes US fintech unicorns viable without international expansion. In 2024, the top three US fintech funding rounds alone — Stripe's $694 million recapitalization, Chime's $300 million Series G extension, and Brex's $235 million credit facility — exceeded the total fintech investment in France, Germany, and the Netherlands combined.

The emerging story is not the US's continued dominance but the rise of three alternative investment centers that are positioning themselves as regulatory arbitrage plays relative to the EU's tightening framework.

Despite the negative impacts of the Iran war at the time of writing, the UAE — specifically the Abu Dhabi Global Market and the Dubai International Financial Centre — has become the most aggressively courted fintech jurisdiction in the world. Both free zones offer regulatory environments that combine genuine supervisory rigor (both are members of IOSCO and maintain regulatory frameworks acceptable to European counterparty due diligence requirements) with speed and flexibility that EU regulation cannot match. The ADGM's AI governance framework explicitly benchmarks itself against the EU AI Act but adopts a principles-based rather than rules-based approach that avoids the specific technical requirements — including conformity assessments and the explainability mandate — that European fintechs cite as their most burdensome obligations. Fintech investment into the UAE reached $2.8 billion in 2023, a 40% increase over 2022, while EU fintech investment fell by 31%.

Singapore has maintained its position as the dominant fintech hub for Southeast Asia while increasingly attracting European and American fintechs seeking an Asia-Pacific base outside China. The Monetary Authority of Singapore's regulatory sandbox approach — which allows fintechs to deploy live products to real customers under temporary regulatory exemptions — provides the kind of real-world testing environment that no EU member state can currently offer without having to complete the full AI Act conformity process. MAS processed forty-seven sandbox applications in 2023, with an average decision time of twelve weeks — a pace that European fintech founders describe with a mixture of envy and frustration.

The United Kingdom, post-Brexit, has made financial services regulatory competitiveness an explicit policy objective in a way that EU membership constrained. The Financial Services and Markets Act 2023 introduced a new secondary objective for UK financial regulators — to facilitate the UK economy's international competitiveness and long-term growth — that has no equivalent in EU regulatory mandates. The FCA has responded with a series of fintech-friendly initiatives: an 'AI Lab' within its regulatory sandbox, a streamlined authorization process for AI-enabled financial products, and explicit guidance that the FCA won't impose EU AI Act-equivalent requirements on UK firms, preferring a 'pro-innovation' principles-based approach. The consequences are beginning to show in investment flows: the UK attracted $10.4 billion in fintech investment in 2023, more than the combined total of France, Germany, Sweden, and the Netherlands.

The Innovation Exodus: European Fintech Founders in Their Own Words

The most telling evidence of how the EU AI Act affects competition isn't found in overall investment numbers, but rather in individual decisions made by founders that these statistics overlook. In early 2025, over 100 CEOs from European startups—including leaders in fintech, healthtech, and AI—signed an open letter urging a halt to enforcing the AI Act. They warned that incomplete regulations and inconsistent national approaches were causing founders to move their main operations outside the EU. Data from Silicon Valley Bank and Atomico support this trend: founders often mention the United Kingdom, UAE, and Singapore as preferred regulatory locations, arguing that the AI Act's requirements for conformity assessments clash fundamentally with the flexible, iterative product development needed for AI-driven fintech innovations.

One German credit scoring startup illustrates the challenge precisely: their credit model is retrained weekly on new data and continuously updated in production. A conformity assessment process that takes months to complete cannot accommodate a model that changes faster than the assessment timeline allows. The choice is between deploying a static model that can be assessed and approved — sacrificing the performance advantage of continuous learning — or deploying the best model outside the EU and a compliant but inferior model within it. Neither option is acceptable for a company competing against US and Asian rivals whose regulatory environments impose no such constraint.

The EU's response to these concerns — the establishment of AI regulatory sandboxes in each member state, designed to allow testing under supervision before full conformity assessment — has been welcomed in principle and criticized in practice. The sandboxes are national rather than EU-wide, meaning a company testing in France cannot automatically deploy in Germany based on its French sandbox approval. The twenty-seven separate national sandbox programs, with different timelines, different entry criteria, and different conditions for graduation to full deployment, replicate at the AI governance layer the regulatory fragmentation that the EU single market was supposed to eliminate.

THE DIGITAL WALLET WARS: A GLOBAL BATTLEFIELD

Digital wallet penetration is the most consequential and least well-understood dimension of the current phase of global fintech competition. The term 'digital wallet' covers an enormous range of products — from Apple Pay's tap-to-pay functionality to WeChat's comprehensive financial super-app to M-Pesa's USSD-based mobile money service — that share the common characteristic of allowing users to transact financially through a digital interface rather than through a physical card or cash. Over 5 billion users projected by 2028 represent a penetration rate among the global adult population that would make digital wallets the primary payment instrument for the majority of humanity.

The competitive dynamics of digital wallet markets differ fundamentally across three distinct global contexts: the high-income markets of North America, Europe, and East Asia, where wallets compete with well-established card payment infrastructure; the middle-income markets of Southeast Asia, Latin America, and the Gulf, where wallets are growing rapidly by improving on an existing but imperfect financial system; and the low-income markets of Sub-Saharan Africa and South Asia, where wallets are building financial infrastructure in areas where the alternative is cash or no financial services at all.

In high-income markets, the digital wallet competition is primarily a battle for the customer's primary payment interface — the moment of checkout, whether physical or digital, in which the customer reaches for their payment instrument. Apple Pay and Google Pay have captured this moment in many high-income markets by integrating payment directly into the devices and operating systems that customers use for everything else. Their market position reflects the distribution advantage described in Chapter One: not a better payment product, but a better payment context. By 2024, Apple Pay was active on over 85% of eligible iPhones in the United States and over 75% in the United Kingdom. Google Pay's dominance in Android markets is less complete but structurally similar.

The competitive response from traditional financial institutions has been PayPal, whose 400-million-user base represents the largest non-platform-embedded wallet in the high-income world, and a series of bank-backed consortia — Zelle in the United States, Wero in Europe, and various national equivalents — that attempt to replicate the convenience of Big Tech wallets while keeping the financial relationship within the banking system. The strategic challenge for these consortia is that they are competing on the turf of companies for whom payments are a distribution feature rather than a core business: Apple and Google can afford to subsidize their payment services in ways that banks cannot, because payments generate data and customer engagement that fuel their primary business models.

In middle-income markets, the most instructive battlegrounds are Southeast Asia and Latin America, where digital wallet penetration has accelerated rapidly but where the competitive landscape has not yet consolidated around dominant players. Southeast Asia's wallet market — contested by Grab, Sea Money, GoPay, and a dozen national and regional competitors —

demonstrates the super-app dynamic that WeChat pioneered in China: the most successful wallets are those embedded in daily-life platforms (ride-hailing, food delivery, e-commerce) rather than those that present themselves primarily as financial products. Grab Financial, which by 2024 provided credit, insurance, and investment products to over 35 million users across eight Southeast Asian countries, built its financial services position on the back of a ride-hailing network that had already established daily habitual engagement before the financial products were introduced.

Latin America's wallet market is defined by the tension between MercadoPago — the financial services arm of the region's dominant e-commerce platform MercadoLibre, which by 2024 processed over $200 billion annually and served 50 million active users — and the expanding financial super-apps built by Nubank, Rappi, and national banking consortia. The region's high financial exclusion rates, its history of currency instability, and its large informal economy create conditions for wallet penetration that closely parallel the Sub-Saharan African mobile money story, but with a more diverse competitive field and more developed digital infrastructure than the African market had at equivalent stages of mobile money development.

AFRICA: THE NEXT WALLET FRONTIER

Sub-Saharan Africa's digital wallet market is in a different phase of development from that of any other major region, and the dynamics shaping its trajectory warrant extended treatment because they will define the financial lives of the largest youth population on earth over the next two decades.

Mobile money — the USSD-based wallet system pioneered by M-Pesa and subsequently replicated across the continent by Airtel Money, MTN Mobile Money, Orange Money, and dozens of national operators — has created a financial infrastructure that bypassed the banking system entirely for much of Sub-Saharan Africa's unbanked population. As of 2024, the GSMA reported over 800 million registered mobile money accounts across Sub-Saharan Africa, with an active user base of approximately 420 million. Transaction volumes exceeded $900 billion annually — a figure that, for context, exceeds the GDP of all but the five largest African economies.

The next phase of African digital wallet development is the smartphone transition. As the penetration of internet-capable devices accelerates — driven by falling device costs, expanding 4G coverage, and the growth of sub-$100 smartphone categories targeted at African consumers — the wallet market is bifurcating. The established USSD-based mobile money operators are racing to build smartphone-native app experiences that retain their existing user bases while offering the richer functionality that app-based wallets can deliver. New entrants — including PalmPay (backed by Chinese investors), Wave (a Senegal-based fintech with a zero-fee remittance model), and OPay (a Nigerian super-app now operating across six African markets) — are building smart-phone-first products that compete with the established operators for the growing segment of African consumers with consistent internet access.

The AI dimension of African wallet development is particularly significant for the financial inclusion argument at the center of this book. Smartphone-native wallets can incorporate AI-powered features — bespoke savings prompts, real-time fraud detection, AI credit assessment, multilingual financial guidance — that USSD-based systems structurally cannot. The wallet that can tell a smallholder farmer in Zambia, in Nyanja rather than English, that her savings are growing toward her stated goal and that this week's harvest price makes this a good moment to pay down her input loan early, is not just a payment instrument. It is a personalized financial advisor, available twenty-four hours a day, calibrated to her specific circumstances, and delivered at a cost that makes it economically viable to serve someone with a balance of fifty dollars.

This is the product that the AI inflection point makes possible. Whether it is the product that gets built — rather than a more sophisticated version of the existing extractive dynamic in which financial products serve their providers' commercial interests at the expense of their users' financial well-being — depends on the design choices, governance frameworks, and commercial incentives that the stakeholders in African fintech create over the next decade.

Wave, which launched its mobile money service in Senegal in 2017–2018, operates on a pricing model that most mobile money operators initially dismissed as commercially unsustainable: dramatically lower fees than incumbent providers. Wave offers free deposits, withdrawals, and bill payments, while charging a flat 1% fee for money transfers, compared with significantly higher transfer fees charged by established telecom-led mobile money systems.

The competitive consequences of this pricing strategy have been striking. Within a few years of launch, Wave became one of the largest mobile money providers in Senegal, rapidly gaining market share from incumbent operators such as Orange Money and Free Money, both of which had been operating in the market for years. Its growth was driven by a combination of lower pricing, a simplified user interface, and a distribution network built around independent agents rather than telecom airtime resellers.

Wave's strategy suggests that the cost structure of mobile money transactions in many African markets reflects incumbent pricing power as much as underlying operating costs. By scaling a lower-fee model across a large user base, Wave has sustained its business through a combination of float income, value-added services, and the network effects that arise when a payment platform reaches large transaction volumes.

At a smaller scale, transaction fees remain economically necessary for many mobile money providers. But Wave's rapid growth indicates that the scale at which lower-fee models become viable may be lower than incumbent operators had assumed, particularly when digital distribution, streamlined onboarding, and strong network effects reduce the marginal cost of each transaction.

TAKEAWAY *Wave illustrates that the most powerful competitive strategy in African mobile money is not product sophistication but price disruption — and that the regulatory environment that permits genuine price competition is a precondition for the consumer welfare benefits that fintech promises.*

EU FINTECH: WHAT IT GETS RIGHT AND WHAT IT RISKS

An honest assessment of the EU's fintech regulatory framework requires acknowledging both what it gets right — which is considerable — and the specific ways in which it risks becoming an obstacle to the innovation it is designed to govern responsibly.

What the EU gets right, above all, is the rights framework. The AI Act's insistence that individuals affected by algorithmic decisions have the right to understand those decisions is more than a compliance burden. It is a substantive protection against the kind of algorithmic harm that the fintech literature documents at length: discriminatory credit decisions, predatory product targeting, and the systematic extraction of value from financially vulnerable consumers who lack the information to protect themselves. The explainability requirement may come at the cost of some accuracy in credit models. It also makes those models legally accountable in a way that their alternatives are not — and accountability is the mechanism through which systematic bias is identified and corrected.

The EU's open banking framework, built through PSD2 and extended through PSD3, is the most technically mature in the world. The UK's equivalent is more developed in practice, but the EU's regulatory architecture for data portability and third-party access is the most thorough legislative framework for consumer data rights in finance that any jurisdiction has produced. When it works — and the implementation record across member states is uneven — it produces exactly the kind of data access levelling that reduces the structural advantage of incumbent platforms and enables competitive innovation.

DORA's operational resilience framework addresses a genuine systemic risk — the concentration of financial services operations among a small number of cloud providers — that no other major jurisdiction has confronted with comparable legislative force. The requirement that critical third-party technology providers be subject to supervisory oversight is, conceptually, the right answer to the cloud concentration problem. The implementation challenges are real but tractable.

What the EU risks, in the aggregate effect of this regulatory architecture, is a fintech investment environment in which the compliance cost of operating in Europe is sufficient to tilt founder and investor decisions toward establishing primary operations elsewhere — in the UK, in Singapore, in the UAE — and serving European customers from outside the EU regulatory perimeter. This isn't merely a competitiveness problem for the European technology industry, though it is that. It is a governance problem: if the most innovative AI financial products are built outside the EU regulatory framework and then distributed into EU markets through the AI Act's extraterritorial provisions, the quality of regulatory oversight over those products will be lower than if they had been built within the regime from the onset. Regulations that drive innovation offshore do not protect European consumers from innovative products. It simply ensures that those products are regulated by someone else, at arm's length, without the access to the development process that domestic supervision would provide.

WHERE FINTECH INVESTMENT TRENDS WILL GO

The structural forces shaping the next phase of global fintech investment are identifiable, even if their precise timing and magnitude are not. Five trends will define where capital flows over the next five years.

The first is the consolidation of the AI-native fintech layer. The companies that will attract the largest investment rounds in 2025 to 2030 will not be the neobanks and payment processors of the first fintech wave — most of which have either achieved sufficient scale to be self-financing or have been acquired by incumbents — but the AI-native financial infrastructure companies that provide the decision intelligence on which next-generation financial products run. Credit assessment AI, fraud detection AI, regulatory compliance AI, and customer intelligence AI are all categories in which venture capital is actively building companies that will eventually be acquired by or partnered with the major financial institutions that lack the internal AI development capability to build these systems organically.

The second trend is the geographic shift of investment toward markets with favorable regulatory environments, combined with large underserved populations. The UAE, Singapore, and the UK will continue to attract disproportionate amounts of fintech capital relative to their population size. Nigeria, Indonesia, and Brazil will attract growing investment as the combination of large unbanked populations, improving digital infrastructure, and maturing regulatory frameworks creates the conditions for fintech development at scale. The EU's share of global fintech investment will continue to fall unless regulatory reform creates a more competitive environment for AI-enabled product development.

The third trend is the embedding of fintech into non-financial platforms at a scale that will make 'fintech' as a distinct category progressively less meaningful. The $7 trillion embedded finance market projected by 2030 — financial services delivered as components of non-financial products and experiences — points to a future in which credit, insurance, payments, and investment are features of agricultural platforms, healthcare systems, logistics networks, and retail networks rather than standalone financial products. The investment implications are significant: the most valuable fintech plays of the next decade may not be recognizable as fintech at all.

The fourth trend is the growing importance of infrastructure investment relative to product investment. The first fintech wave was primarily a product wave: new financial products built on existing infrastructure. The second wave — the AI wave — requires new infrastructure: AI model training infrastructure, real-time data processing infrastructure, identity verification infrastructure, and regulatory reporting infrastructure that can handle the volume and complexity of AI-mediated financial decisions. The companies building this infrastructure — Plaid, Stripe, Marqeta, TrueLayer, and dozens of less visible API-layer companies — will capture a growing share of the value created by the fintech revolution even as the consumer-facing product companies attract more of the public attention.

The fifth trend is the consolidation of digital wallet markets around super-app models that integrate financial services with daily life platforms. The standalone digital wallet — an app whose primary function is payment — will be progressively absorbed into broader digital life platforms in the same way that standalone music players were absorbed into smartphones. The wallets that survive as independent products will be those that have built the daily engagement, data richness, and service breadth that make them truly indispensable rather than merely convenient.

The Regulatory Trilemma: Innovation, Safety, and Competitiveness

Financial regulation appears to face a persistent trade-off between innovation, safety, and international competitiveness. Different jurisdictions weigh these objectives differently. The EU AI Act leans more heavily toward safety and rights protection; the US model has generally leaned more toward market dynamism; and jurisdictions such as Singapore and the UAE have relied more heavily on sandbox-based experimentation under supervision.

The EU AI Act implicitly prioritizes safety and rights protection over competitiveness. The result is the most comprehensive rights framework for AI in finance in the world, at the cost of competitive disadvantage in the speed and flexibility of AI product deployment. The US approach implicitly prioritizes competitiveness over safety, producing the world's most dynamic fintech network alongside the widest documented gaps in consumer protection and systemic oversight. Singapore and the UAE attempt to balance all three through regulatory sandbox architectures that permit innovation under supervision — an approach that works well for smaller markets with concentrated supervisory capacity but scales less readily to the complexity of larger financial systems.

The author's own research on the regulatory quality complement to fintech development — which found that better-governed fintech markets produce larger development dividends — is relevant to this trilemma, but with an important qualification. The regulation that produced the largest fintech-development dividend in the China panel study was fintech-specific regulation designed for the technology, not financial regulation adapted from pre-digital frameworks. The EU AI Act is, structurally, the latter: a regulatory framework designed by legislators who understood AI as a technology to be governed, rather than by practitioners who understood AI as a financial tool to be made to work safely. The difference in perspective produces a regulation that is highly protective and not very enabling — the opposite balance from what the China data suggests produces the largest development dividend.

The most important regulatory reform the EU could make — short of substantively revising the AI Act, which the political investment in its passage makes impractical in the near term — is to build a proper EU-wide AI sandbox architecture that allows companies to test products across multiple member states under a single regulatory authorization, with a clear pathway from sandbox testing to full deployment that does not require repeating the conformity process for each member state. This reform would not resolve the fundamental trilemma. It would reduce the speed constraint that fintech founders identify as their most damaging competitive disadvantage.
Sources: KPMG Pulse of Fintech H2 2023; Juniper Research Digital Wallets 2024–2028; MAS FinTech Festival Singapore 2023 Industry Survey; Atomico State of European Tech 2024; author's analysis extending Muganyi et al. (2022) regulatory quality findings

These three chapters share a common diagnosis: the AI inflection point is not being navigated with the governance architecture it requires. Central banks are supervising AI-mediated financial systems with tools designed for human-mediated ones. Geopolitical actors are exploiting the dollar's structural vulnerabilities in ways that the AI and fintech revolution has made technically feasible for the first time. And the most ambitious regulatory framework for AI in finance — the EU AI Act — is producing unintended competitive consequences that may ultimately reduce the quality of governance over AI financial products, rather than improve it, by pushing development offshore.

There is no reason to lose hope based on these findings. Central banks are investing in supervisory technology at a pace and seriousness that has no precedent in their history. The dollar's structural foundations — the depth of US capital markets, the network effects of global trade invoicing, the safe asset demand for US Treasuries — are more durable than the de-dollarization rhetoric acknowledges. And the EU's regulatory framework, whatever its competitive costs, is establishing the rights and accountability norms that the global AI financial system needs, even if Europe itself does not capture the full commercial benefit of doing so.

What they do counsel is urgency — a recognition that the window in which governance frameworks can shape the AI financial system's trajectory is narrowing as the system's architecture becomes progressively more entrenched. The choices being made now — about central bank AI governance, about reserve currency alternatives, about the balance between innovation and accountability in AI financial products — will be far harder to revise in ten years than they are today. The infrastructure of human possibility that the opening chapter began with — the twelve digits and a hash key that connected a woman in Filabusi to the global financial system — is being rebuilt at a scale and complexity that the architects of the first fintech wave could not have anticipated. Getting its governance right, now, is the most important work in financial policy of our time.

CHAPTER FIFTEEN

The AI Infrastructure Divide

*Compute, Sovereignty, and the
Architecture of Financial Power*

*The most serious form of financial exclusion in the coming
decade will not be the lack of a bank account. It will be the lack of the
computational infrastructure on which the financial system runs.*

When I started working on this book, I spent a lot of time reading. Papers, reports, working notes, anything that could help me understand how finance and artificial intelligence were beginning to overlap in meaningful ways. The process was intense, but it was also made easier by the tools available today. AI systems could summarize long academic papers, highlight key arguments, and help me move through material much faster than I would have otherwise. It felt like I had access to something powerful. I noticed that when I was using these tools in Africa, the experience was not always consistent. The network would drop. Responses would take longer to load. Sometimes they would not load at all. Certain features were unavailable, even though they were part of the same platform. It was easy to dismiss at first. Connectivity issues are not new. Anyone working in this environment learns to adapt. You refresh the page. You try again. You move between networks.

Then I travelled to Europe.

The difference was immediate. The same tools behaved differently. Responses were faster, more stable, more reliable. Features that had been missing before were suddenly available without any effort. The experience felt seamless in a way that I had not realized was possible until I saw it for myself. That contrast forced me to pause. For many people in developed markets, access to digital services is assumed. When something does not work, it is seen as a temporary inconvenience. A streaming service might not carry a certain show in one region. A simple workaround solves the problem. It is frustrating, but it is not structural. What I was experiencing felt different. This was not about content. It was about capability.

The ability to access and use AI tools consistently depends on infrastructure. Network stability, bandwidth, data centers, compute availability. These are not abstract concepts. They shape what is possible on a day-to-day basis. They determine who can use these systems effectively and who cannot. When those foundations are uneven, the outcomes become uneven as well. The conversation around artificial intelligence often focuses on models, applications, and use cases. Less attention is given to the infrastructure that supports all of it. Without that infrastructure, the technology does not disappear, but access to it becomes fragmented. That fragmentation has consequences. If researchers, developers, and businesses in certain regions cannot rely on stable access to these tools, their ability to participate in the broader ecosystem is limited. The gap is not just about knowledge or talent. It is about the environment in which that talent operates.

The first fourteen chapters of this book have examined the AI inflection point largely through the lens of products, platforms, and regulatory frameworks — the visible layer of the fintech revolution. This chapter looks beneath that layer, at something more fundamental and weightier: the physical and computational infrastructure on which the entire AI-driven financial system depends. The argument is straightforward and, once stated, difficult to dismiss. If the future of financial services is AI-native — if credit decisions, risk assessments, fraud detection, regulatory compliance, and customer engagement are all mediated by machine learning models — then the geography of AI infrastructure becomes the geography of financial power. And that geography is profoundly unequal.

To understand why this matters, consider what an AI-native financial system actually requires. It requires compute — the processing power to train and run machine learning models at scale. It requires data centers — the physical facilities that house the servers on which those models operate. It requires training data — the vast, structured datasets from which models learn to make financial decisions. It requires connectivity — the high-speed, low-latency networks that allow AI systems to operate in real time. And it requires talent — the engineers, researchers, and data scientists who build, maintain, and improve these systems. Every one of these requirements is concentrated, to a degree that should alarm anyone who cares about financial equity, in a handful of countries.

The numbers are stark. They should be uncomfortable reading for anyone who takes the financial inclusion agenda seriously. The United States accounts for approximately 33% of global data center capacity. China accounts for roughly 16%. Europe collectively holds about 17%. The rest of the world — home to more than five billion people and containing the vast majority of the populations that fintech is supposed to serve — shares the remainder, with most of it concentrated in a few hub nations: Singapore, Japan, Australia, India, and the UAE. Sub-Saharan Africa, which contains some of the world's fastest growing fintech markets, accounts for less than one percent of global data center capacity. Latin America accounts for roughly three percent. Southeast Asia, excluding Singapore, accounts for less than five percent.

This concentration is not accidental. Data centers require reliable electricity — enormous quantities of it — and the regions with the weakest AI infrastructure are frequently the regions with the least reliable power grids. They require capital investment measured in billions of dollars per facility. They require a regulatory environment that permits the storage and processing of financial data under conditions acceptable to international financial institutions. They require cooling systems, physical security, and connectivity to submarine cable networks. These are not constraints that a well-designed app can overcome. They are structural features of the global economy that reproduce themselves with each new generation of technology.

The semiconductor supply chain compounds the asymmetry. The advanced chips required to train and run large AI models — the GPUs and TPUs on which the entire AI financial infrastructure depends — are designed mainly in the United States, manufactured predominantly in Taiwan and South Korea, and subject to export controls that the United States has used with increasing aggressiveness to limit China's access to the most advanced processors. No African country manufactures semiconductors. No Latin American country designs the advanced chips required for AI model training. The dependence is not just commercial. It is architectural: the computational substrate on which AI-native finance runs is controlled, at every stage of its production, by a small number of nations and an even smaller number of companies.

THE THREE LAYERS OF AI FINANCIAL POWER

A useful framework for understanding how this infrastructure asymmetry translates into financial power distinguishes three layers, each building on the one below.

At the base sits infrastructure. This comprises the physical and computational foundation: the chips, the data centers, the cloud platforms, the high-speed networks, and the energy systems that power them. Control of this layer determines who can build AI systems at all. It is the most concentrated layer and the hardest to replicate, because its requirements — capital, technology, energy, and talent — are precisely the resources that the Global South possesses in the least abundance.

Above it sits the model layer. This comprises the foundation models, the specialized financial AI models, and the AI agents that are trained on the infrastructure layer and deployed to make financial decisions. Control of this layer determines the logic of financial decision-making: whose risk models are used, whose credit assessment algorithms are trusted, whose fraud detection systems set the standard. The companies building these models — the major US and Chinese AI labs, the large financial institutions with internal AI research capacity — are overwhelmingly headquartered in the same countries that control the infrastructure layer.

The top layer — the one the user touches — is applications. This comprises the consumer-facing fintech products — the digital wallets, the lending platforms, the insurance products, the investment tools — that are built on top of the model layer. This is the layer that the financial inclusion literature has focused on almost exclusively, because it is the layer that the end user interacts with. It's also the layer with the lowest barriers to entry and the most geographic diversity. A startup in Lagos or Jakarta or São Paulo can build a fintech application. But if that application runs on cloud infrastructure hosted in Virginia, uses a credit scoring model trained in California, and depends on an AI framework maintained in Mountain View, then the startup's apparent independence is an illusion. It is operationally dependent on an infrastructure stack it does not control, cannot inspect, and has no meaningful ability to influence.

This three-layer structure — infrastructure, models, applications — is the new architecture of financial power. And it reproduces, in digital form, a pattern that development economists have documented in other contexts for decades: a pattern in which the Global South provides the markets, the data, and the users, while the Global North provides the infrastructure, the intelligence, and the profits.

AI FINANCIAL SOVEREIGNTY

The concept of financial sovereignty has traditionally referred to a nation's ability to control its own monetary policy, regulate its own financial institutions, and manage its own currency. The AI inflection point adds a new dimension: AI financial sovereignty, which refers to a nation's ability to control — or at least meaningfully influence — the computational infrastructure, the data governance frameworks, and the AI models on which its financial system depends.

India's approach to this challenge is instructive. The India Stack — comprising Aadhaar, UPI, and the India Account Aggregator framework — represents the most deliberate attempt by a large developing economy to build sovereign digital financial infra-structure. By maintaining public ownership of the identity layer, payment rails, and data-sharing framework, India has laid the foundation for domestic fintech companies to build on without ceding architectural control to foreign platforms. The Data Empowerment and Protection Architecture extends this logic to

AI by establishing principles for how financial data can be collected, shared, and used to train AI models, with the explicit goal of ensuring that Indian financial data generates value within India rather than being extracted for model training abroad.

Africa's situation is more fragmented and more precarious. The African Continental Free Trade Area has created, in principle, the conditions for a pan-African digital financial infrastructure. In practice, the continent's AI infrastructure is overwhelmingly dependent on hyperscale cloud providers headquartered in the United States, with African data frequently stored and processed in European data centers under European data governance frameworks. The African Union's AI strategy, adopted in 2024, acknowledges this dependency and calls for investment in African data center capacity, but the capital requirements — estimated at tens of billions of dollars — dwarf the budgets available to most African governments. Rwanda and Kenya have emerged as early movers, with Kigali positioning itself as a regional data center hub and Nairobi attracting investment from both Western and Chinese cloud providers. But the gap between aspiration and capacity remains enormous.

Latin America faces a distinct version of the same challenge. Brazil's Pix instant payment system — which processed more than 42 billion transactions in 2023 — demonstrates that the region can build world-class financial infrastructure. But Pix is a payments rail, not an AI infrastructure. The AI models that Brazilian fintech companies use for credit scoring, fraud detection, and customer segmentation are overwhelmingly built on US cloud platforms, trained on frameworks developed in US research labs, and optimized for data patterns that may not fully reflect the economic realities of Brazilian consumers. Mexico, Colombia, and Argentina face similar dependencies, compounded by smaller domestic markets and less developed regulatory frameworks for AI governance.

Southeast Asia presents the most dynamic and contested landscape. Singapore functions as the region's AI financial hub, with data center capacity, regulatory sophistication, and talent density that rival any Western financial center. But Singapore's 5.9 million population means its domestic market cannot sustain the kind of sovereign AI infrastructure that India or China can build on the basis of domestic demand alone. Indonesia, the region's largest economy with 275 million people, has the market scale but not yet the infrastructure: its data center capacity is growing rapidly but remains a fraction of Singapore's, and its AI talent pipeline, while expanding, cannot yet support the development of indigenous financial AI models at the frontier. The result is a regional architecture in which Singapore provides the infrastructure and regulatory overlay, while the larger Southeast Asian economies provide the markets and the data — a division that benefits the region as a whole but concentrates architectural control in its smallest member.

THE NEW FINANCIAL COLONIALISM

It would be an overstatement — but not a large one — to describe the emerging AI financial structure as a new form of financial colonialism. The historical pattern of colonialism involved the extraction of raw materials from colonized territories, their transformation into finished goods in the colonizing power, and the sale of those finished goods back to the colonized market at prices that captured the value of transformation. The AI financial structure replicates this pattern with data as the raw material, AI models as the transformation process, and fintech products as the finished goods sold back to the markets from which the data was extracted.

A farmer in Malawi whose mobile money transactions are processed by a platform running on AWS infrastructure in Frankfurt, using a credit scoring model trained in San Francisco on data patterns derived from Kenyan and Nigerian users, is participating in an economic relationship that extracts more value than it delivers to her local economy. The transaction fees flow to the platform. The cloud computing fees flow to Amazon. The model development costs were amortized across a global user base that she is part of but has no voice in. The regulatory framework under which her data is processed was designed in a jurisdiction in which she has no representation. She has access to a financial product — which is of real value — but she has no access to the infrastructure that makes it possible, no ownership of the data that makes it work, and no influence over the algorithmic logic that determines her creditworthiness.

This isn't an argument against the products. The products work. The preceding chapters of this book provide compelling evidence supporting their effectiveness. Furthermore, the discussion highlights that the underlying infrastructure supporting these products has distributional implications which remain insufficiently examined within the financial inclusion literature.

TOWARD AI INFRASTRUCTURE EQUITY

What would an equitable AI financial infrastructure look like? The question is easier to pose than to answer, but several principles can be identified.

First, the development of regional AI compute capacity should be treated as a financial infrastructure investment, not as a technology policy afterthought. Just as the construction of payment rails, clearing systems, and banking networks was understood in the twentieth century as a prerequisite for financial development, the construction of AI compute infrastructure should be understood in the twenty-first century as a prerequisite for AI-native financial development. Development finance institutions — the World Bank, the African Development Bank, the Asian Development Bank — should treat data center investment and AI training infrastructure with the same seriousness they have historically given to payments infrastructure.

Second, data governance frameworks should be designed to ensure that the financial data generated in a jurisdiction creates value within that jurisdiction. This does not require data localization in its crudest form — which can impede cross-border financial services — but it does require frameworks that give nations meaningful control over how their citizens' financial data is used for AI model training, and that ensure the models trained on that data are available to domestic financial institutions on terms that do not perpetuate dependency.

Third, the open-source AI movement — which has already produced foundation models of remarkable capability — should be actively supported by development institutions as a counterweight to the proprietary model architectures controlled by a handful of US and Chinese companies. An open-source financial AI model, trained on diverse global data and available for adaptation by local financial institutions, would represent a genuine alternative to the extractive model that currently dominates.

The future of financial inclusion, this chapter argued, depends not only on access to financial products but on access to the AI infrastructure that makes those products possible. The twelve digits and a hash key that connected a woman in Filabusi to the formal financial system were revolutionary because they bypassed the infrastructure barriers — the branch networks, the documentation requirements, the physical distance — that had excluded her. The AI inflection point is creating a new set of infrastructure barriers, just as serious and considerably harder to see. Addressing them isn't an optional extension of the financial inclusion agenda. It is the next frontier on the financial inclusion agenda.

CHAPTER SIXTEEN

Financial Autonomy

From Account Ownership to Algorithmic Autonomy

The question that defined the first era of financial inclusion was: does this person have a bank account? The question that will define the next era is: does this person have an AI that works for them?

I remember the first time I watched Garth Jennings' adaptation of *The Hitchhiker's Guide to the Galaxy* on the big screen. It was sometime in the mid-2000s, long after Douglas Adams had already embedded the story in popular culture, but the moment still carried a strange sense of anticipation. In the film, a supercomputer named Deep Thought spends 7.5 million years calculating the answer to what is described as the *Ultimate Question of Life, the Universe, and Everything*. The entire galaxy waits for the result. Civilizations rise and fall while the machine quietly works through its calculations. The expectation is that the answer will be something profound — a philosophical revelation that will finally make sense of existence.

When the calculation is complete, the machine calmly delivers its conclusion.

The answer is 42.

The absurdity of the moment is the point. Deep Thought explains that the real problem was never the calculation itself, the problem was that nobody had properly defined the question. The universe had spent millions of years waiting for an answer. Perhaps to a question it didn't actually understand. I often find myself thinking about that scene when reading discussions about artificial intelligence in finance.

There is an enormous amount of excitement — and anxiety — about what AI will do to the financial system. We debate whether algorithms will replace human bankers, whether automated credit decisions will make finance fairer or more dangerous, and whether markets will become more efficient or more fragile. The conversation is intense and to many quite urgent. For others there is a surprising certainty about outcomes that remain deeply uncertain.

But beneath all of that discussion sits a quieter problem.

Are we actually asking the right questions?

Much of the current debate treats artificial intelligence as a tool — a faster calculator, a more sophisticated risk model, a better fraud detection system. In this view, AI improves finance in the same way that spreadsheets once improved accounting: by making existing processes more efficient. But that framing may be far too narrow.

The deeper transformation lies not in automation, but in autonomy.

For most of its history, finance has been a human decision system supported by technology. Algorithms provided information, but the final judgement remained with people — the loan officer assessing a borrower, the trader evaluating a market signal, the regulator deciding whether a risk model was credible.

Artificial intelligence begins to change that relationship. Modern AI systems do not simply calculate outcomes; they increasingly generate decisions, often through internal processes that even their designers struggle to fully interpret. Credit scoring models update themselves continuously as new data arrives. Fraud detection systems learn patterns that no human analyst could identify manually. Algorithmic trading systems operate at speeds where human intervention becomes functionally impossible.

In other words, parts of the financial system are beginning to move toward algorithmic autonomy — systems that do not merely assist human judgement but operate with a degree of independent decision-making power. This raises a difficult question that economists, regulators, and technologists are only beginning to confront.

What happens when the infrastructure of finance is no longer just digital, but autonomous?

The challenge is not simply technical. It is epistemological. Human financial systems have always depended on the assumption that someone, somewhere, understands the logic behind important decisions. A loan is approved because certain criteria are met. A risk model behaves in a predictable way. A regulator can trace how a failure occurred.

Autonomous AI systems complicate that assumption. Some of the most powerful machine-learning models function as black boxes, producing accurate predictions without easily interpretable explanations. Their internal reasoning can be statistically valid but cognitively opaque — patterns that make sense to the mathematics of the model but not to the human mind observing it.

In that sense, the analogy with Deep Thought becomes unexpectedly relevant.

The financial system may increasingly rely on machines capable of producing answers — credit decisions, investment strategies, risk forecasts — whose internal reasoning is not fully accessible to the humans who depend on them. The system may work extremely well. It may even outperform traditional human judgement.

But we may not always understand why.

The shift that this chapter describes is not speculative. It is already underway. AI financial agents — software systems that act autonomously on behalf of a user to manage money, optimize spending, negotiate financial products, allocate savings, and execute investment strategies — are moving from research prototypes to commercial deployment. The first generation of these agents is relatively primitive: automated savings tools that round up purchases, robo-advisors that allocate investment portfolios according to pre-set rules, and chatbots that help users navigate insurance claims. But the trajectory is clear, and the capabilities are advancing at a pace that outstrips the ability of regulatory frameworks — or inclusion metrics — to keep up.

FROM HUMAN-MANAGED TO AGENT-MANAGED FINANCE

My aunt in Filabusi was one of the first people I saw truly benefit from financial access. Not in a theoretical sense, but in a way that changed how she lived day to day. She did not have a smartphone. She did not need one. With a cracked handset and a USSD code, she could send and receive money, check her balance, and begin to build something that resembled financial stability. It was simple. It was powerful.

Access alone created visibility. It allowed her to see what she had, what she could save, and what she could plan for. Now I find myself asking a different question. What happens when financial systems do more than record transactions and begin to understand them? I ponder about what her life would look like today with the tools that are starting to emerge. Not as an abstract idea, but as something practical. Something she could wake up to.

A simple dashboard on a smartphone that reflects her day's work. Sales recorded automatically. Income categorized without her having to think about it. Portions of that income allocated quietly into savings, into school fees, into inputs for the next planting season. No forms. No manual tracking. No need to sit down at the end of the week and calculate what is left. I think about the cooperative meetings she attended every Friday. Women gathering to pool their resources, to decide together how funds would be allocated. There was value in that. It built trust, accountability, and community. At the same time, it required time, coordination, and physical presence. What if those decisions could happen continuously, guided by systems that understood the needs of the group and the patterns of their income? What if the structure of that cooperative could exist without the friction that came with it?

I reflect on the journeys she had to make when something went wrong. Taking a bus into the city, carrying a sample of a withered tomato or a dead chicken, waiting to be told what had caused the problem. Hours lost. Money spent. Uncertainty carried all the way there and back. What if that knowledge was available instantly? What if the same device that handled her finances could also analyze her crops, detect early signs of disease, and suggest what to do before the loss became irreversible? These are not distant ideas anymore. The pieces are already coming together. What they point to is something deeper than access.

They point to autonomy.

The ability to operate, to decide, and to adapt without constant dependence on external systems or intermediaries. The ability to move from reacting to circumstances to shaping them. For someone like my aunt, that shift would not just improve her business. It would change how she relates to risk, to opportunity, and to time itself. It would create space for planning, for growth, for decisions that are not constrained by immediate uncertainty.

And it would not stop with her. The effects would move through the community. Other women in the cooperative. Families connected through remittances. Small networks of trust that could now operate with greater confidence and coordination.

The fundamental shift is from human-managed finance to agent-managed finance. In the human-managed model — which is the model that every existing financial inclusion metric assumes — a person opens an account, makes deposits, decides how to save, chooses whether to borrow, selects an insurance product, and manages the day-to-day flow of money through their financial life. Financial literacy is important because the person is the decision-maker. Product design matters because the person must be able to understand and use the product. Trust matters because the person must choose to engage.

In the agent-managed model, the locus of decision-making shifts. The person sets goals, preferences, and constraints — save for school fees, keep a minimum balance for emergencies, find the cheapest insurance that covers hospitalization — and the AI agent executes. It monitors income flows and automatically adjusts saving rates. It compares financial products across providers in real time and switches when better terms become available. It negotiates interest rates, identifies billing errors, detects fraudulent charges, and optimizes the timing of payments to minimize fees and maximize returns. The person is still the principal. The agent is the fiduciary. But the skills required of the person are fundamentally different: not financial literacy in the traditional sense, but the ability to articulate goals and evaluate whether the agent is serving them well.

The implications for how we measure financial inclusion are difficult to overstate. Consider the traditional metrics. Bank account ownership: in an agent-managed world, the relevant question is not whether a person has an account but whether their agent has access to the full range of financial services on their behalf. A person with one account but with an effective AI agent may be more financially included — in every meaningful sense — than a person with five accounts and no agent. Credit access: in an agent-managed world, the agent negotiates credit terms, compares providers, and manages repayment schedules. The relevant inclusion metric is not whether credit is available but whether the agent can access it on competitive terms. Digital wallet adoption: in an agent-managed world, the wallet is the agent's interface, not the user's. A person who never opens their wallet app but whose agent manages their money effectively is not disengaged. They are well served.

NEW METRICS FOR A NEW ERA

If the old metrics are becoming obsolete, what should replace them? Five dimensions of financial inclusion seem appropriate to the agent-managed era, and they share a common logic: they measure capability rather than access.

The first is algorithmic financial autonomy: how far a person's AI agent can act independently and effectively on their behalf across the full range of financial services. This measures not whether the person has access to products, but whether their agent has the capability, the data, and the authorization to optimize their financial life.

The second is AI advisory access: whether the person has access to an AI financial agent at all, and if so, the quality and capability of that agent. In the same way that access to a competent human financial advisor has historically been a privilege of wealth, access to a capable AI financial agent may become the defining axis of financial inequality. The risk is real: if the best AI agents are available only to premium customers, while lower-income users are served by stripped-down versions with limited capability, the agent-managed era could widen the very inequalities it has the potential to close.

The third is data ownership: how much control a person has over the financial data that their agent uses to make decisions on their behalf. An agent that runs on proprietary data held by a platform the person cannot leave is not serving the person's autonomy, even if it is serving their immediate financial interests. True financial inclusion in the agent era requires portable data — data that the person owns, that moves with them between providers, and that their agent can use on any platform.

The fourth is financial decision augmentation: the degree to which the AI agent enhances rather than replaces the person's financial understanding. The best agents will not just execute. They will explain; why they made a particular choice, what alternatives they considered, what risks they identified, and what the person should understand about their own financial situation. An agent that manages money well but leaves the person no more financially capable than they were before has solved a problem without building capacity. An agent that manages money well and makes the person more financially literate in the process has achieved something truly game-changing.

Finally, agent-to-agent financial interoperability: the degree to which AI agents can transact, negotiate, and cooperate across platforms, providers, and jurisdictions. In a fully agent-managed financial system, the critical infrastructure is not the payment rail or the banking network but the protocol by which agents communicate. If agents cannot interoperate — if a person's savings agent cannot communicate with their insurance agent, or if agents on different platforms cannot negotiate directly — then the agent-managed era will reproduce the fragmentation and friction of the human-managed era in algorithmic form.

THE RISKS OF AGENT-MANAGED FINANCE

The risks of agent-managed finance deserve candid treatment because they are substantial and they're not the risks that the current regulatory conversation is focused on.

The first risk is exclusion through algorithmic bias. If AI agents make financial decisions based on patterns learned from historical data, and that historical data reflects the systematic exclusion of women, ethnic minorities, rural populations, and other marginalized groups, then the agents will reproduce that exclusion at scale and at speed. The difference from human bias is not moral — algorithmic bias is no less unjust than human bias — but operational: an AI agent that discriminates does so consistently, at volume, and often invisibly.

Then there is the over-automation problem. There is a meaningful difference between an agent that helps a person manage their finances and an agent that manages their finances for them while they disengage entirely. Financial engagement is not purely instrumental — a cost to be minimized by delegating to a machine. It is constitutive of economic agency: the sense that one understands, controls, and directs one's own economic life. An AI agent that is so effective that the person never thinks about money has solved the efficiency problem at the cost of the agency problem. Designing agents that augment human engagement rather than replace it is one of the most important design challenges of the next decade.

The most insidious risk, however, is dependence on proprietary AI systems. If a person's entire financial life is managed by an agent built and operated by a single platform, the person's financial autonomy is an illusion. They're not financially independent. They are financially dependent on a company. Switching costs in agent-managed finance could dwarf the switching costs in today's financial system, because moving from one agent to another means transferring not just accounts but the accumulated intelligence — the learned preferences, the optimization history, the relationship context — that makes the agent effective. Without portability requirements, agent-managed finance could produce the most locked-in customer relationships in the history of financial services.

The conclusion follows from the argument, and it can be stated simply. The future question of financial inclusion is not whether someone has a bank account, a digital wallet, or access to credit. It is whether someone has an AI agent that works for them — acting in their interest, with their interests as its objective function, their data under their control, and their understanding as its secondary output.

Building that future requires not only better technology but better metrics, better regulation, and a fundamental rethinking of what it means to be financially included in an age of intelligent machines.

CHAPTER SEVENTEEN

The AI–Fintech Trade-offs

Efficiency, Fairness, and the Architecture of Compromise

Every new AI-powered financial product sits somewhere along a set of tensions that cannot be resolved — only managed. The societies that manage them wisely will prosper. The societies that pretend the tensions do not exist will be governed by them.

I s it possible that simply having an iPhone could cause you to be charged more for the same products or services than if you had an Android device?

It sounds like something abstract until you see it happen. A friend of mine in India once pointed out that she was being charged different prices on the same app depending on the device she used. When she logged in with an Android phone, the prices were slightly lower. When she used an iPhone, there was a noticeable mark-up. Nothing dramatic. Just enough to raise a question. Once she saw it, she could not ignore it.

This is not as strange as it first appears. Digital systems are built to observe behavior and respond to it. Device type, location, timing, and spending patterns all become signals. These signals are not just collected. They shape outcomes. They influence what is shown, what is offered, and sometimes what is charged. From the perspective of the platform, the logic is straightforward. If different users respond differently to price, then prices can be adjusted. The system becomes more efficient. Revenue improves. It works as intended.

From the perspective of the user, it feels different.

There is a sense that something is happening beneath the surface. The price is no longer just a number. It becomes a result, shaped by factors that are not always visible or understood. This is where the idea of trade-offs begins to matter. Companies rely on data to refine how they operate. They want to improve performance, increase engagement, and grow sustainably. At the same time, there is an expectation that this process does not cross into something that feels unfair or exploitative. The boundary is not always clear. Artificial intelligence changes how this plays out. These systems do not pause to question intent. They learn from patterns. If the data shows that certain users are willing to pay more, the system can act on that information. It does not need to justify the outcome. It follows the structure it has been given. What might once have been a conscious decision becomes embedded behavior. The trade-off is no longer debated in a meeting. It is carried out continuously, at scale, and often without visibility.

What my friend experienced is a small example, but it points to something much larger. As AI becomes more embedded in financial systems, the same logic can extend into areas that matter far more. Pricing is only one part of it. Access to credit, risk assessments, and financial opportunities can all be shaped in similar ways. These systems are inherently built to optimize. The question is how that optimization is defined, and whether the outcomes it produces align with what people consider fair, reasonable, and acceptable.

The preceding chapters have documented in considerable detail the disruptive potential of AI in financial services and the governance challenges that transformation creates. This chapter takes a different approach. Rather than examining specific technologies, products, or regulatory frameworks, it examines the structural tensions — the trade-offs — that the convergence of AI and finance creates, and that every jurisdiction, every institution, and every product designer must navigate. These tensions are not problems to be solved. They are features of the system, inherent in the architecture of AI-mediated finance, and the quality of governance depends on how honestly and skillfully they are managed.

EFFICIENCY VERSUS FAIRNESS

The first and most fundamental tension is between efficiency and fairness. AI systems are, at their core, optimization engines. They are designed to identify patterns, minimize costs, maximize accuracy, and allocate resources with a precision that human decision-makers cannot match. In financial services, this optimization capability is truly game-changing. An AI credit underwriting system can process a loan application in seconds, drawing on thousands of data points to assess risk with a granularity that no human loan officer could achieve. AI fraud detection systems can identify suspicious patterns across millions of transactions in real time. AI insurance pricing can calibrate premiums to individual risk profiles with actuarial precision that was previously impossible.

But optimization is not neutral. An AI system that maximizes predictive accuracy for credit default will, inevitably, learn to use features that correlate with protected characteristics — race, gender, geography, ethnicity — even if those features are not explicitly included in the model. A credit scoring system trained on historical lending data will learn that applicants from certain postcodes default more frequently and will penalize those applicants accordingly — reproducing the redlining patterns that the data reflects, at scale and without human intervention. An insurance pricing model that calibrates premiums to individual health risk may achieve actuarial efficiency while pricing the sickest and most vulnerable people out of coverage entirely. The system is efficient. Whether it is fair depends on whose definition of fairness you use.

This tension cannot be resolved by technical means alone. Algorithmic fairness researchers have identified multiple mathematical definitions of fairness — demographic parity, equalized odds, calibration, individual fairness — and have proven that several of these definitions are mutually incompatible. A credit scoring model cannot simultaneously achieve equal approval rates across demographic groups and equal predictive accuracy within each group, except in the limiting case where the groups have identical underlying distributions. In the real world, they do not. The choice between definitions of fairness is not a technical choice. It is a political and moral choice — one that the designers of AI financial systems are implicitly making every time they select an objective function and a training methodology.

PERSONALISATION VERSUS PRIVACY

Personalization and privacy pull in opposite directions, and the tension between them sits at the heart of the AI-native financial promise. The promise of AI-native financial services — the promise that drives much of the investment and enthusiasm documented in this book — is personalization at scale: financial products calibrated to the specific circumstances, preferences, and needs of each individual user. Delivering on that promise requires data. Enormous quantities of data. Not just transaction histories and credit records, but behavioral data, location data, social network data, communication patterns, spending habits, health information, employment trajectories, and dozens of other data categories that, taken together, constitute a comprehensive map of a person's economic life.

The more data an AI financial system has about a user, the better it can serve that user. This is not a hypothesis. It is a mathematical property of machine learning systems: model performance improves with data quality and quantity, subject to diminishing returns but with no theoretical maximum. A credit scoring model with access to a user's transaction history, mobile phone usage patterns, social connections, and geolocation data will outperform a model with access to transaction history alone. An AI financial advisor with access to a person's full economic picture — income, expenses, debts, savings, insurance, health status, family obligations — will provide better advice than one working from partial information.

The privacy cost of this improvement is substantial and asymmetric. The users who stand to benefit most from AI-powered personalization — low-income users with thin credit files, users in developing markets with limited formal financial histories, users whose financial lives are complex and irregular — are also the users who are most vulnerable to data exploitation. They have less bargaining power over how their data is used. They have fewer meaningful alternatives if they object to a platform's data practices. And they face greater consequences from data breaches, because their financial margins are thinner and their recourse options fewer.

Case analysis lays bare the tension concretely. AI credit underwriting in markets like Kenya and Indonesia has expanded credit access to millions of previously unbanked users by scoring creditworthiness on the basis of mobile phone data — call patterns, airtime purchases, app usage. The inclusion gains are genuine. But the data collection required to achieve them means that a person's mobile phone — the most personal device they own — becomes a permanent audit trail for financial assessment, with no clear boundary between what is collected for credit scoring and what might be used for other purposes entirely.

AUTOMATION VERSUS HUMAN AGENCY

The automation-agency tension is the one that keeps financial ethicists awake at night. If AI manages finances better than humans — and the evidence suggests that in many specific tasks, it already does — then the rational response is to automate as much as possible. An AI system that optimizes savings rates, minimizes fees, detects fraud, rebalances investments, and negotiates insurance premiums will outperform the average human on every one of these tasks. The efficiency gains are real. The welfare improvements are measurable.

But the displacement of human judgement from financial decision-making has costs that efficiency metrics do not capture. Financial decisions are not purely instrumental. They are exercises of agency — expressions of a person's values, priorities, and vision for their own life. The decision to save for a child's education rather than a home renovation, to accept a lower investment return for a socially responsible portfolio, to lend money to a family member at a loss — these are decisions that reflect who a person is, not just what they can afford. An AI system that optimizes for financial outcomes without regard for these human dimensions of financial life may improve the balance sheet while impoverishing the experience.

The challenge is particularly acute in developing markets, where financial engagement is not just a convenience but a vehicle for economic and social empowerment. The woman in Filabusi whose story opened this book was not just accessing a financial system. She was exercising economic agency — watching her savings grow, making choices about how to use them, participating in a financial community with other women in her cooperative. An AI agent that managed her finances more efficiently but removed her from the process would have delivered better financial outcomes at the cost of the agency that made those outcomes meaningful. The task for designers is not to choose between automation and agency but to build systems that deliver the efficiency of the former without sacrificing the empowerment of the latter.

GLOBAL PLATFORMS VERSUS LOCAL FINANCIAL ECOSYSTEMS

Scale versus sovereignty is the tension that matters most at the macroeconomic level. AI-driven fintech platforms benefit enormously from scale. A credit scoring model trained on hundreds of millions of users across dozens of markets will outperform a model trained on a single market. A fraud detection system that monitors global transaction patterns will catch threats that a local system would miss. A financial AI that operates in 50 countries can amortize its development costs across all of them, offering services at a price point that no domestic competitor can match.

The efficiency advantages of global platforms are real and substantial. But they come at a cost to local financial networks. When a global platform captures the majority of digital financial transactions in a developing market, the domestic financial institutions that might have served those customers lose the data, the revenue, and the learning that would have allowed them to develop their own capabilities. The local financial ecosystem does not simply lose market share. It loses the capacity to develop because the data and the relationships that drive AI improvement are being captured by a foreign platform with no obligation to reinvest them locally.

Central bank oversight tools represent one response to this tension. Several central banks — the Bank of England, the Monetary Authority of Singapore, and the Reserve Bank of India — are building AI-powered supervisory systems that can monitor the activities of global platforms operating within their jurisdictions. These systems use natural language processing to analyze platform disclosures, machine learning to detect patterns of non-compliance, and network analysis to map the interconnections between global platforms and local financial institutions. They are necessary but not sufficient: they address the supervisory challenge without addressing the underlying structural issue of domestic capability loss.

THE AI FINTECH TRADEOFF MATRIX

These four tensions — efficiency versus fairness, personalization versus privacy, automation versus agency, global platforms versus local ecosystems — are not independent. They interact, reinforce each other, and create compound challenges that no single policy instrument can address. A framework that this chapter proposes for handling them is the AI Fintech Trade-off Matrix: a diagnostic tool that maps any AI-powered financial product or policy along each of the four tensions simultaneously.

The matrix operates on a simple principle. Every AI fintech innovation represents a set of choices — explicit or implicit — about where it sits on each tension. An AI credit scoring system that uses alternative data sources to serve underbanked populations has chosen a specific position on the personalization-privacy axis (more data, less privacy) and on the efficiency-fairness axis (potentially fairer, if the alternative data reduces demographic bias, or potentially less fair, if it introduces new biases). An automated investment platform that removes human advisors from the process has chosen a specific position on the automation-agency axis. A global fintech platform that enters a developing market has chosen a position on the global-local axis.

Making these choices explicit — requiring innovators, regulators, and policymakers to articulate where their products and policies sit on each tension, and to justify those positions — does not resolve the trade-offs. But it transforms the governance conversation from a reactive exercise in damage control to a proactive exercise in design. It forces the question that too much of the fintech discourse has avoided: not whether AI should enter finance, because it already has, but how societies choose to manage the trade-offs that its presence creates.

The real challenge of the AI-fintech era is not technological. It is architectural. It is the challenge of building financial systems that are simultaneously efficient and fair, personalized and private, automated and human, global and local. No system will achieve all of these simultaneously. Every system will involve compromises. The quality of governance will be measured not by the lack of trade-offs but by the honesty and wisdom with which they are acknowledged, negotiated, and managed.

CHAPTER EIGHTEEN

2040: A Scenario for AI-Driven Finance

*Intelligence, Inclusion, and the
Architecture of What Comes Next*

*The future of financial inclusion will not be determined
by access to banking — but by access to intelligence.*

On nights when I can't sleep, I sometimes find myself wondering what the story of AI-driven finance will look like a decade from now. Not the abstract story told in research papers or conference panels, but the ordinary, everyday version — the one lived by people who may never read a single page about fintech or artificial intelligence.

Recently, my two-year-old gave me a glimpse of how strange that future might look.

We were in Cape Town, and she watched me pay for a latte by tapping my phone on a small NFC terminal. The transaction took less than a second. No card, no cash, no signature — just a brief vibration from the phone and the quiet confirmation that the payment had gone through. To her, it was a curious little magic trick. What made it even stranger was that only a few weeks earlier she had watched me pay for vegetables in Zimbabwe using cash. Notes counted out by hand. Change returned across a market stall. Two entirely different financial worlds separated by nothing more than a plane flight.

Since then, she has developed her own theory of how the system works. She regularly steals my phone, walks over to the refrigerator, taps the device against the door, and then looks at me expectantly — apparently convinced that this is the correct procedure for authorizing an unlimited snack supply. In her mind, the logic is perfectly sound: tap the object, receive the reward. Watching this small misunderstanding play out has made me wonder what the real financial system will look like for her generation. Will smartphones remain the primary interface with money? Will payments migrate to watches, glasses, or some other wearable technology? Or will the entire process become so embedded in the environment that the act of "paying" disappears altogether — replaced by systems that operate quietly in the background of everyday life?

That question is the premise of the somewhat ill-advised exercise that follows. Predicting the future of finance is famously dangerous. Entire libraries are filled with confident forecasts that were obsolete within a few years of publication. Yet if this book has highlighted one consistent theme, it is that the trajectory of financial technology points toward a system that becomes progressively less visible. The most successful innovations in finance are rarely the ones that draw attention to themselves. They are the ones that disappear into the routines of daily life — systems so seamless that people stop noticing them entirely.

This final chapter, therefore, departs from the analytical mode that has defined the rest of the book. It is not an argument, nor a piece of empirical analysis. It is a scenario — a disciplined attempt to imagine what the financial world might look like in 2040 if the forces examined in the previous seventeen chapters continue along their current trajectories.

Scenarios are not predictions. They are tools for thinking about consequences. At their best, they illuminate decisions that are being made today whose effects will only become visible years from now. The scenario that follows is offered in that spirit.

A DAY IN 2040

Amara wakes in Kigali at six in the morning. Before she opens her eyes, her financial agent — an AI system she has named Keza, after her grandmother — has already completed its overnight cycle. It has reconciled yesterday's transactions across her three income streams: the salary from her position as a logistics coordinator at the East African Trade Corridor Authority, the rental income from the small apartment she owns in Nyamirambo, and the returns from her share in a cooperative agricultural investment fund that pools capital from 12,000 smallholder farmers across Rwanda, Uganda, and Tanzania.

Keza has also, overnight, renegotiated Amara's health insurance premium. The insurer's AI agent proposed a 4% increase based on regional health trend data. Keza counter-offered with Amara's personal health metrics — shared with explicit consent under Rwanda's AI Data Sovereignty Act of 2032 — which show fitness levels and preventive care engagement well above the regional average. The two agents settled on a 1.2% increase, logged the negotiation record to the blockchain-anchored audit trail that Rwandan regulation requires, and presented Amara with a one-sentence summary when she checks her morning dashboard: "Health insurance renewed. Saved you 14,200 RWF versus the standard rate."

This is unremarkable. It is Tuesday. This is what Tuesday looks like in Kigali in 2040. The reader may find it implausible. It is not. Every component of the system Amara is using either exists in prototype today or is a straightforward extension of systems that do.

By eight o'clock, Amara is at her desk. Her work involves coordinating shipments along the Kigali–Dar es Salaam corridor, a route that carries agricultural products, manufactured goods, and raw materials through three countries. The financial infrastructure of this corridor has changed beyond recognition in the past fifteen years. Payments between Rwandan exporters and Tanzanian buyers settle in real time through the Pan-African Instant Settlement Network — a system built on infrastructure that the African Union and the African Development Bank spent seven years constructing, explicitly to reduce the continent's dependence on Western correspondent banking networks. The settlement system runs on AI-optimized routing that selects the cheapest and fastest payment path for each transaction, across a network of central bank digital currencies that are interoperable by design.

Amara does not think about any of this. She thinks about logistics. The financial infrastructure is invisible to her — as it should be. Her AI agent handles the financial dimensions of each shipment: currency conversion, insurance, customs duties, and payment timing. When a shipment is delayed by weather in Dodoma, Keza automatically triggers the parametric insurance contract, which pays out within 11 minutes based on satellite weather verification. No claim form. No adjuster. No dispute. The contract was written in machine-readable code, the triggering conditions were verified by an AI that monitors East African weather data in real time, and the payout was executed by the insurer's settlement agent.

Amara receives a notification: "Weather delay payout: 380,000 TZS credited to corridor operations account."

THE ADVANCED HUBS

Amara's experience is representative of the advanced AI financial hubs — the jurisdictions that invested early in sovereign AI infrastructure, built robust governance frameworks, and achieved what the development literature now calls full-spectrum financial inclusion: not merely access to accounts and products, but access to AI agents, data sovereignty, and computational infrastructure.

Singapore remains the world's most sophisticated AI financial center. Its regulatory sandbox, now in its fourth generation, processes more than 2,000 AI financial product applications annually. The Monetary Authority of Singapore operates what is widely regarded as the most advanced AI supervisory system in the world — a network of machine learning models that monitors the activities of more than 400 financial institutions in real time, detecting emerging risks, compliance failures, and systemic vulnerabilities before they materialize.

India's AI financial infrastructure, built on the foundation of Aadhaar, UPI, and the Account Aggregator framework, now serves 1.3 billion people through a network of AI financial agents that are interoperable by regulatory mandate. The National AI Financial Agent Standard, introduced in 2035, requires every AI agent operating in India's financial system to be compatible with every other agent, to be portable between providers, and to explain its decisions in any of India's twenty-two scheduled languages. The standard is imperfect. Compliance is uneven, particularly among smaller providers. But it has created a financial ecosystem in which agent-to-agent interoperability is the norm rather than the exception, and in which a farmer in Bihar has access to the same quality of AI financial guidance as a software engineer in Bangalore — if not always the same range of products.

The UAE has positioned itself as the global hub for AI-driven Islamic finance, building on two decades of regulatory innovation to create a model that permits AI agents to operate within sharia-compliant parameters. Dubai's AI Financial Free Zone, established in 2033, has attracted more than 600 AI fintech companies and manages assets exceeding two trillion dollars. The zone's regulatory approach — permissive on innovation, strict on consumer protection and data governance — has become a model for jurisdictions across the Middle East and North Africa.

THE EMERGING ADAPTERS

That is the optimistic picture. It isn't the only one. The emerging adapters — markets that are adopting AI-driven finance rapidly but haven't yet built the sovereign infrastructure to do so on their own terms — face a different set of challenges.

Nigeria, now Africa's largest economy by a substantial margin, has achieved near-universal access to AI financial agents through a partnership model that the country's central bank describes as "guided dependency." The major AI financial platforms operating in Nigeria are built on infrastructure hosted partly in the country's own data centers — the Lagos AI Compute Centre, opened in 2034 with African Development Bank financing — and partly on cloud infrastructure operated by global providers under Nigerian data governance rules. The model works: Nigerian consumers have access to sophisticated AI financial services. But the dependency on foreign model architectures means that the algorithmic logic of Nigerian financial decision-making is still substantially shaped by training data and design choices made in Silicon Valley and Shenzhen.

Brazil's Pix ecosystem has evolved into one of the world's most advanced real-time financial networks, processing more than 200 billion transactions annually. The country's AI financial agent market is competitive and diverse, with more than forty providers offering agents tailored to Brazil's complex tax system, seasonal agricultural income patterns, and informal economy. But Brazil's data governance framework remains contested: the tension between the country's ambitious AI strategy and its underfunded data protection authority has created a regulatory environment that is innovative in principle but inconsistent in practice.

Indonesia's 290 million people represent one of the world's largest AI financial markets, and the country's agent adoption rate has grown faster than any other in Southeast Asia. The challenge is distribution: the gap between Java's sophisticated urban financial infrastructure and the eastern provinces' limited connectivity means that AI financial agents work well for 180 million Indonesians and poorly for the rest. The country's ambitious "AI for the Archipelago" program, launched in 2036, is building satellite-linked AI financial access points in remote areas, but progress is slower than the program's architects anticipated.

THE INFRASTRUCTURE GAP

Here is where the scenario turns dark. The most troubling feature of the 2040 financial landscape is not the regions that have succeeded or the regions that are adapting. It is the regions where the AI infrastructure gap has hardened into a structural barrier that no amount of application-layer innovation can overcome.

Parts of Sub-Saharan Africa, Central Asia, and the Pacific Islands remain dependent on AI financial services that are hosted, trained, and governed entirely abroad. The users in these regions have access to financial products — often good products, delivered by platforms that are genuinely trying to serve them well. But they have no ownership of the data those products generate, no influence over the algorithms that govern their financial lives, and no domestic infrastructure on which alternative systems could be built. They are consumers of AI-driven finance, not participants in it.

The consequences are measurable. The regions with sovereign AI financial infrastructure — India, Singapore, the UAE, Rwanda, and a handful of others that invested early — show financial development outcomes that are 30 to 50% stronger than comparable regions without that infrastructure, controlling for income, connectivity, and education. The mechanism is not mysterious: sovereign infrastructure enables domestic AI model development, which enables products calibrated to local economic realities, which enables deeper financial engagement, which drives the development dividend this book's second chapter documented using the China data. Without sovereign infrastructure, the cycle does not start.

This is the scenario. It isn't inevitable. It is conditional — conditional on choices being made now, in the mid-2020s, about infrastructure investment, data governance, regulatory design, and the architecture of AI financial systems. The choices are still open. They will not remain open indefinitely.

BEYOND INCLUSION

The title of this book is Beyond Inclusion. It was chosen because these eighteen chapters argue that the concept of financial inclusion — as traditionally understood, as traditionally measured, and as traditionally pursued — is no longer sufficient for the world that the AI inflection point is creating.

Financial inclusion, in its first-wave formulation, meant access: getting people into the financial system. It meant bank accounts, mobile wallets, digital payments, and the infrastructure that made them possible. That formulation was correct for its era, and its achievements were extraordinary. More than a billion people entered the formal financial system in a single decade. Mobile money transformed the economic lives of hundreds of millions of people across Sub-Saharan Africa and South Asia. Digital payments compressed the cost of financial transactions in ways that benefited entire economies. The evidence for these gains, documented in the early chapters of this book, is rigorous and compelling.

But access is not enough. The evidence is equally clear on this point. The access-engagement gap — the chasm between holding an account and actively using the financial system to build resilience, accumulate wealth, and exercise economic agency — remains one of the most stubborn features of the global financial landscape. Hundreds of millions of people who nominally have access to financial services do not use them in ways that generate the development dividends the inclusion agenda promised.

The AI inflection point transforms the terms of this challenge. It creates the possibility of financial systems that understand their users as individuals — systems that can deliver personalized guidance, calibrate products to specific circumstances, detect vulnerability before it becomes a crisis, and empower people to make financial decisions that serve their own goals rather than the goals of the institutions that serve them. It creates the possibility of AI agents that act as genuine financial fiduciaries for the billions of people who have never had access to competent financial advice. It creates the possibility of a financial system that isn't merely inclusive but intelligent — one that does more than open the door but helps people find their way through it.

It also creates the possibility of a financial system that is more unequal than anything that has come before. A system in which the best AI serves the wealthiest users, the best infrastructure serves the most connected nations, the best data governance serves the most powerful jurisdictions, and the efficiency gains of AI-native finance are captured by the platforms and the countries that control the computational substrate on which everything runs. This is not a speculative risk. The trends documented in this book — the concentration of AI infrastructure, the extractive data architectures, the regulatory asymmetries, the platform dependencies — are already producing this outcome in measurable ways.

The central thesis of this book is that the future of financial inclusion will not be determined by access to banking but by access to intelligence. Intelligence understood in the broadest sense: the intelligence embedded in AI systems that mediate financial decisions, the intelligence embedded in data governance frameworks that determine who benefits from the data the financial system generates, the intelligence embedded in regulatory architectures that balance innovation with accountability, and the intelligence embedded in infrastructure investments that determine which populations participate in the AI-native financial system and which merely consume its outputs.

The twelve digits and a hash key that connected a woman in Filabusi to the formal financial system represented a revolution in access. The next revolution — the one this book has tried to map, analyze, and argue for — is a revolution in intelligence. It is a revolution that is already underway, that is moving faster than the governance architectures designed to shape it, and that will determine, more than any other force in the global economy, who prospers and who is left behind in the decades ahead.

Getting it right is the work of our time.

LIST OF ABBREVIATIONS

ABCD	Artificial Intelligence, Blockchain, Cloud Computing, and Data Analytics
ADGM	Abu Dhabi Global Market
AI	Artificial Intelligence
AML	Anti-Money Laundering
API	Application Programming Interface
ATM	Automated Teller Machine
AWS	Amazon Web Services
BaaS	Banking as a Service
BBVA	Banco Bilbao Vizcaya Argentaria
BCB	Banco Central do Brasil
BIS	Bank for International Settlements
BNPL	Buy Now, Pay Later
BoE	Bank of England
BRICS	Brazil, Russia, India, China, and South Africa
CBDC	Central Bank Digital Currency
CEO	Chief Executive Officer
CFPB	Consumer Financial Protection Bureau

CFTC	Commodity Futures Trading Commission
CGAP	Consultative Group to Assist the Poor
CIPS	Cross-Border Interbank Payment System
CNY	Chinese Yuan
COBOL	Common Business-Oriented Language
DBS	Development Bank of Singapore
DeFi	Decentralized Finance
DEX	Decentralized Exchange
DIFC	Dubai International Financial Centre
DORA	Digital Operational Resilience Act
EBA	European Banking Authority
ECB	European Central Bank
ESG	Environmental, Social, and Governance
EU	European Union
FCA	Financial Conduct Authority
FSB	Financial Stability Board
FTT	FTX token
FTX	FTX Trading Ltd
FX	Foreign exchange
GANDALF	Google, Apple, Netflix, DBS, Amazon, LinkedIn, Facebook
GDP	Gross Domestic Product
GDPR	General Data Protection Regulation
GFDI	Global Fintech Development Index

GPU	Graphics Processing Unit
GSMA	Global System for Mobile Communications Association
GST	Goods and Services Tax
HSBC	Hongkong and Shanghai Banking Corporation
IEA	International Energy Agency
IMF	International Monetary Fund
IOSCO	International Organization of Securities Commissions
IPO	Initial Public Offering
KYC	Know Your Customer
LMIC	Low- and Middle-Income Country
MAS	Monetary Authority of Singapore
MiCA	Markets in Crypto-Assets Regulation
ML	Machine Learning
NIST	National Institute of Standards and Technology
NLP	Natural Language Processing
OCC	Office of the Comptroller of the Currency
PSD2	Revised Payment Services Directive
PSD3	Third Payment Services Directive
RMF	Risk Management Framework
SDG	Sustainable Development Goal
SEC	Securities and Exchange Commission
SPFS	System for Transfer of Financial Messages
SWIFT	Society for Worldwide Interbank Financial Telecommunication

TSB	Trustee Savings Bank
UK	United Kingdom
UPI	Unified Payments Interface
US	United States
USD	United States Dollar
USDC	USD Coin
USDT	USD Tether
USSD	Unstructured Supplementary Service Data

GLOSSARY

Agent banking A model in which licensed financial institutions engage third-party retail outlets to offer basic banking services on their behalf, extending reach beyond traditional branch networks.

Algorithmic bias Systematic and repeatable errors in computer systems that create unfair outcomes, such as privileging or discriminating against particular groups, often resulting from biased training data or flawed model design.

Autonomous finance Financial systems in which AI agents independently execute transactions, manage portfolios, and make credit decisions with minimal or no human intervention.

Banking as a Service (BaaS) A model in which licensed banks offer their regulated infrastructure—payments, lending, compliance—via APIs, enabling non-bank companies to embed financial products in their own platforms.

Big tech Large technology companies—such as Google, Apple, Amazon, Meta, and Microsoft—that have expanded into financial services, leveraging existing user bases and data ecosystems.

Central Bank Digital Currency (CBDC) A digital form of a country's fiat currency issued and backed by the central bank, designed to function as legal tender alongside physical cash.

Contagion The spread of financial distress from one institution, market, or country to others through interconnected exposures, payment systems, or investor behavior.

Credit scoring (alternative) The use of non-traditional data sources—mobile phone usage, utility payments, social media activity—to assess creditworthiness, particularly for individuals without formal credit histories.

Decentralized finance (DeFi) A category of financial applications built on blockchain networks that seek to replicate traditional financial services—lending, trading, insurance—without centralized intermediaries.

Digital financial inclusion The process of ensuring that individuals and businesses, particularly those historically excluded from formal banking, can access and use affordable, appropriate digital financial services.

Embedded finance The integration of financial services—payments, lending, insurance—directly into non-financial platforms, applications, or ecosystems, so that users access them without visiting a bank or financial institution.

Fintech Technology-driven innovation in financial services, encompassing start-ups, products, and business models that use software, data analytics, and digital platforms to deliver or enhance financial products.

Fintech–AI inflection point The thesis advanced in this book: that the convergence of fintech and artificial intelligence represents a structural turning point in global financial development, shifting the system from access-driven inclusion toward intelligence-driven autonomy.

Financial development The process by which financial institutions, instruments, and markets grow in depth, breadth, and efficiency, supporting economic growth and broader access to financial services.

Financial stability A condition in which the financial system—institutions, markets, and infrastructure—operates without serious disruption, absorbing shocks without amplifying them into broader economic crises.

GPU compute Processing power provided by graphics processing units, originally designed for rendering images but now essential for training and running large-scale AI and machine learning models.

Interoperability The ability of different financial systems, platforms, or technologies to communicate, exchange data, and process transactions across institutional and national boundaries.

Inverted U risk curve A pattern identified by the author's research describing the non-linear relationship between AI adoption intensity and bank risk. Initial adoption increases risk, deeper adoption achieves a threshold and integration leads to mitigated risk.

Know Your Customer (KYC) Regulatory requirements obliging financial institutions to verify the identity, suitability, and risks associated with maintaining a business relationship with a customer.

Leapfrog paradox The finding, identified in the author's panel study, that countries which leapfrogged traditional banking infrastructure via fintech may face diminishing—or even reversing—financial development returns as fintech matures.

M-Pesa A mobile money transfer and payments service launched in Kenya in 2007 by Safaricom, widely credited as a foundational example of mobile-led financial inclusion in the developing world.

Macro-prudential regulation Regulatory policies aimed at safeguarding the stability of the financial system as a whole, rather than individual institutions, by monitoring and mitigating systemic risks.

Mobile money A technology that allows users to store, send, and receive money using a mobile phone, typically without requiring a traditional bank account.

Model risk The risk of adverse consequences arising from decisions based on incorrect or misused model outputs, including AI and machine learning models used in credit scoring, trading, or risk management.

Open banking A regulatory and technological framework requiring banks to share customer data—with customer consent—via APIs with authorized third-party providers, fostering competition and innovation.

Regulatory arbitrage The practice of structuring activities to exploit differences between regulatory regimes, jurisdictions, or classifications in order to reduce compliance costs or avoid restrictions.

Regulatory sandbox A controlled testing environment established by a financial regulator that allows firms to trial innovative products, services, or business models under relaxed regulatory requirements for a limited period.

Regtech Technology-driven solutions designed to help financial institutions meet regulatory compliance requirements more efficiently, including automated reporting, identity verification, and transaction monitoring.

Stablecoin A type of cryptocurrency designed to maintain a stable value by pegging it to an external reference, typically a fiat currency such as the US dollar, or a basket of assets.

Super-app A mobile application that combines multiple services—messaging, payments, shopping, ride-hailing, lending—into a single platform, creating a self-contained digital ecosystem. WeChat and Grab are prominent examples.

Systemic risk The risk that the failure or distress of one financial institution, market segment, or infrastructure component triggers cascading failures across the wider financial system.

Techfin A term distinguishing technology companies that move into financial services (e.g., Alibaba, Tencent) from financial institutions that adopt technology (fintech), emphasizing the different regulatory and competitive dynamics involved.

Unstructured Supplementary Service Data (USSD) A communications protocol used by mobile networks that allows real-time, session-based interaction via simple text menus on basic handsets, widely used for mobile money services in developing countries.

REFERENCES AND FURTHER READING

Full citations for all empirical claims, statistics, and quotations used in this book are provided below, organized by chapter. Where the author's own primary research is the source, the published reference and methodology are given in full. All third-party statistics are sourced from the references cited; readers are encouraged to consult the sources for full methodological context.

Primary Research by the Author

Muganyi, T., Yan, L., Yin, Y., Sun, H., Gong, X. & Taghizadeh-Hesary, F. (2022). Fintech, regtech, and financial development: evidence from China. Financial Innovation 8(29). https://doi.org/10.1186/s40854-021-00313-6

Muganyi, T. (2023). Do Fintech AI Developments Exacerbate Bank Risk in the Eurozone? MSc Dissertation, Warwick Business School / Bank of England Programme. Accepted for oral presentation at the Economics of Financial Technology Conference, University of Edinburgh, June 2024.

Muganyi, T., Yan, L. & Sun, H. (2021). Green finance, fintech and environmental protection: evidence from China. Environmental Science and Ecotechnology 7, 100107.

References and further reading

Abu Dhabi Global Market (2023). Guidance for Digital Securities Activities and Virtual Assets. Abu Dhabi: ADGM.

African Continental Free Trade Area Secretariat (2023). Protocol on Digital Trade (under negotiation). Accra: AfCFTA.

African Development Bank (2023). African Economic Outlook 2023: Mobilizing Private Sector Financing for Climate and Green Growth. Abidjan: AfDB.

African Union (2024). Continental Artificial Intelligence Strategy. Addis Ababa: African Union Commission.

Ant Group / Alibaba Group (2020). Prospectus for Initial Public Offering. Hong Kong: Hong Kong Stock Exchange.

Atomico (2024). State of European Tech 2024. London: Atomico.

Bain & Company (2022). Embedded Finance: What It Takes to Prosper in the New Value Chain. Boston: Bain & Company.

Banco Central do Brasil (2023). Pix: A New Brazilian Instant Payments Ecosystem. Brasília: BCB.

Banco Central do Brasil (2024). Pix Statistics: Annual Review 2023. Brasília: BCB.

Bank for International Settlements (2020). Central bank digital currencies: foundational principles and core features. BIS Papers No. 1. Basel: BIS.

Bank for International Settlements (2021). Big tech in finance: opportunities and risks. BIS Annual Economic Report. Basel: BIS.

Bank for International Settlements (2022). Triennial Central Bank Survey: OTC Foreign Exchange Turnover in April 2022. Basel: BIS.

Bank for International Settlements (2024). Annual Economic Report, Chapter III: AI and the Economy. Basel: BIS.

Bank for International Settlements (2024). Intelligent financial system: how AI is transforming finance. BIS Working Papers No. 1194. Basel: BIS.

Bank of England (2022). Financial Stability Report, December 2022. London: Bank of England.

Bank of England (2023). Financial Stability Report, December 2023. London: Bank of England.

Bank of England (2024). Machine Learning in UK Financial Services. London: Bank of England / FCA.

BFA Global & Women's World Banking (2022). Women's Financial Inclusion in the Digital Age. New York: Women's World Banking.

BIS Innovation Hub (2022). Project mBridge: Connecting Economies Through CBDC. Basel: BIS.

BIS Innovation Hub (2023). Project Aurora: The Power of Data, Technology and Collaboration to Combat Money Laundering Across Institutions and Borders. Basel: BIS.

BRICS (2023). XV BRICS Summit Johannesburg II Declaration. Johannesburg: BRICS.

CB Insights (2024). State of Fintech Report 2024. New York: CB Insights.

CGAP (2025). Beyond Account Access: Converting Financial Inclusion into Resilience. CGAP Blog Series on Global Findex 2025. Washington DC: CGAP.

Chen, B. et al. (2022). Fintech and Financial Risks of Systemically Important Commercial Banks in China. Sustainability 14(10).

Chouldechova, A. (2017). Fair prediction with disparate impact: a study of bias in recidivism prediction instruments. Big Data 5(2): 153–163.

Cross-Border Interbank Payment System (CIPS). System Overview and Transaction Statistics. Shanghai: CIPS.

DBS Bank Limited (2015–2024). Annual Reports. Singapore: DBS Group Holdings.

Demirgüç-Kunt, A., Klapper, L., Singer, D. & Ansar, S. (2022). The Global Findex Database 2021: Financial Inclusion, Digital Payments, and Resilience in the Age of COVID-19. Washington DC: World Bank.

Dubai International Financial Centre (2023). DIFC Innovation Hub Annual Review 2023. Dubai: DIFC.

Eichengreen, B. (2011). Exorbitant Privilege: The Rise and Fall of the Dollar. Oxford: Oxford University Press.

European Banking Authority (2020). EBA Report on Big Data and Advanced Analytics. EBA/REP/2020/01. Paris: EBA.

European Central Bank (2023). Digital Euro: Progress Report and Third Report on a Digital Euro. Frankfurt: ECB.

European Central Bank (2023). The Rise of Artificial Intelligence: Benefits and Risks for Financial Stability. ECB Financial Stability Review, November 2023. Frankfurt: ECB.

European Commission (2020). A Digital Finance Strategy for the EU. COM/2020/591 final. Brussels: European Commission.

European Parliament (2024). Regulation (EU) 2024/1689: The Artificial Intelligence Act. Official Journal of the European Union.

Federal Reserve Bank of Atlanta. GDPNow Model Documentation. Atlanta: Federal Reserve Bank of Atlanta.

Federal Reserve Board (2023). The International Role of the US Dollar. FEDS Notes. Washington DC: Federal Reserve.

Financial Conduct Authority (2024). AI Update. London: FCA.

Financial Conduct Authority (2024). Regulatory Sandbox. London: FCA.

Financial Stability Board (2017). Artificial Intelligence and Machine Learning in Financial Services: Market Developments and Financial Stability Implications. Basel: FSB.

Financial Stability Board (2023). Global Regulatory Framework for Crypto-asset Activities. Basel: FSB.

Financial Stability Board (2024). Annual Report 2024. Basel: FSB.

GSMA (2023). State of the Industry Report on Mobile Money 2023. London: GSMA.

GSMA Connected Women (2023). The Mobile Gender Gap Report 2023. London: GSMA.

HM Treasury (2023). Financial Services and Markets Act 2023: Factsheet on International Competitiveness. London: HM Treasury.

International Energy Agency (2021). Net Zero by 2050: A Roadmap for the Global Energy Sector. Paris: IEA.

International Energy Agency (2024). Electricity 2024: Analysis and Forecast to 2026 — Data Centre and AI Energy Demand. Paris: IEA.

International Monetary Fund (2023). Global Financial Stability Report, Chapter 3: Generative Artificial Intelligence in Finance. Washington DC: IMF.

International Monetary Fund (2024). Currency Composition of Official Foreign Exchange Reserves (COFER), Q4 2024. Washington DC: IMF.

Juniper Research (2024). Digital Wallets: Market Forecasts, Emerging Trends & Vendor Analysis 2024–2028. Basingstoke: Juniper Research.

Kleinberg, J., Mullainathan, S. & Raghavan, M. (2017). Inherent Trade-Offs in the Fair Determination of Risk Scores. Proceedings of Innovations in Theoretical Computer Science (ITCS 2017).

KPMG (2024). Pulse of Fintech H2 2023. London: KPMG International.

McKinsey & Company (2022). Europe's Fintech Opportunity. McKinsey Global Institute.

Monetary Authority of Singapore (2022). Fairness, Ethics, Accountability and Transparency (FEAT) Principles Assessment Methodology. Singapore: MAS.

Monetary Authority of Singapore (2023). Project MindForge: Generative AI Risk Framework for Financial Institutions. Singapore: MAS.

Monetary Authority of Singapore (2024). Annual Report 2023/24: Fintech and Innovation. Singapore: MAS.

Murinde, V., Rizopoulos, E. & Zachariadis, M. (2022). The impact of the FinTech revolution on the future of banking. International Review of Financial Analysis 81, 102103.

National Institute of Standards and Technology (2023). Artificial Intelligence Risk Management Framework: AI RMF 1.0. Gaithersburg: NIST.

National Payments Corporation of India (2024). UPI Ecosystem Statistics: Annual Report 2023. New Delhi: NPCI.

Nubank (2021). S-1 Registration Statement. Washington DC: US Securities and Exchange Commission.

Ozili, P. K. (2018). Impact of Digital Finance on Financial Inclusion and Stability. Borsa Istanbul Review 18(4): 329–340.

Peking University Digital Finance Research Centre (2019). China Digital Financial Inclusion Index 2019. Beijing: PKU DFIRC.

People's Bank of China (2023). Progress of Research and Development of e-CNY in China. Beijing: PBC.

People's Bank of China (2024). RMB Internationalisation Report 2024. Beijing: PBC.

Pesaran, M. H. (2015). Testing weak cross-sectional dependence in large panels. Econometric Reviews 34(6–10): 1089–1117.

Philippon, T. (2016). The FinTech Opportunity. NBER Working Paper 22476. Cambridge MA: National Bureau of Economic Research.

Prasad, E. (2021). The Future of Money: How the Digital Revolution is Transforming Currencies and Finance. Cambridge MA: Harvard University Press.

Pula Advisors (2023). Annual Impact Report. Nairobi: Pula.

Reserve Bank of India (2023). Report on Trend and Progress of Banking in India. Mumbai: RBI.

Revolut (2025). Annual Report and Accounts 2024. London: Revolut Ltd.

Sahay, R. et al. (2015). Rethinking Financial Deepening: Stability and Growth in Emerging Markets. IMF Staff Discussion Notes 2015/008. Washington DC: IMF.

Semiconductor Industry Association (2024). SIA Factbook 2024. Washington DC: SIA.

Setser, B. (2024). Follow the Money: Blog and Policy Analysis. New York: Council on Foreign Relations.

Suri, T. & Jack, W. (2016). The long-run poverty and gender impacts of mobile money. Science 354(6317): 1288–1292.

Sveriges Riksbank (2022). The Riksbank's Work on a Digital Krona. Stockholm: Sveriges Riksbank.

SWIFT (2024). RMB Tracker: Monthly Reporting and Statistics on Renminbi Progress Towards Becoming an International Currency. Brussels: SWIFT.

Synergy Research Group (2024). Worldwide Data Center Market Trends. Reno: Synergy Research Group.

Tencent Holdings Limited (2024). Annual Report 2024. Shenzhen: Tencent.

US Department of the Treasury (2024). Report on Foreign Holdings of US Treasury Securities. Washington DC: US Treasury.

Wang, R., Liu, J. & Luo, H. (2021). Fintech development and bank risk taking in China. European Journal of Finance 27(4–5): 397–418.

World Bank (2023). Digital Economy for Africa (DE4A) Initiative: Progress Report 2023. Washington DC: World Bank.

World Bank (2023). Migration and Development Brief 38. Washington DC: World Bank.

World Bank (2025). The Global Findex Database 2025. Washington DC: World Bank.

ABOUT THE AUTHOR

Dr Tadiwanashe Muganyi is an economist and financial policy scholar whose work sits at the intersection of financial technology, development economics, and financial regulation. He holds a **PhD in Industrial Economics** and an **MSc in Global Central Banking and Financial Regulation** from **Warwick Business School**, a program delivered in collaboration with the **Bank of England**.

His research has been published in peer-reviewed journals including *Financial Innovation* and *Environmental Science and Ecotechnology* and has been cited in academic and policy discussions internationally. His research on the relationship between fintech-driven artificial intelligence development and bank risk in the Eurozone formed the basis of his MSc thesis at Warwick Business School, completed with distinction.

Alongside his academic work, Dr Muganyi has engaged with financial policy and practice across both African and European markets. This experience informs his research with a perspective grounded in the real-world challenges of financial exclusion that fintech innovation seeks to address.

He is the founding partner of **Nexcelia Consulting**, where he leads research and advisory work focused on the intersection of technology, finance, and economic development in emerging markets.

Beyond Inclusion: Fintech's AI Inflection Point is his first book.

INDEX

Wirecard, *Chapter 9*

Y

Yu'e Bao, *Chapters 1, 2*

Z

Z-scores (bank risk), *Chapter 9*

Zhang, Allen, *Chapter 1*